ML

A practical introduction to the new logic symbols

Second edition

**For all my family,
especially Hazel,
whose dependency requires no definition**

Some well chosen words for all my colleagues affected by this standard:

A person gets from a symbol the meaning he puts into it, and what is one man's comfort and inspiration is another's jest and scorn.

Justice Robert Jackson, *West Virginia State Board v Barnette* (1943)

Meanings change amazingly. When people get accustomed to horrors, these form the foundation for good style.

Boris Pasternak, *Safe Conduct* (1931)

The meaning of things lies not in the things themselves but in our attitude towards them.

Saint-Exupéry, *The Wisdom of the Sands* (1948)

Finally a quotation for all users of the new standard who may be tempted at some time to use an over-complex symbol:

The ability to simplify means to eliminate the unnecessary so that the necessary may speak.

Hans Hofmann, *Search for the Real* (1967)

A practical introduction to the new logic symbols

Second edition

Ian Kampel, C.Eng, MIERE

Butterwortns
London · Boston · Durban · Singapore · Sydney · Toronto · Wellington

First published 1985
Second edition 1986

© I. J. Kampel, 1985

British Library Cataloguing in Publication Data

Kampel, I. J.
   A practical introduction to the new logic symbols.—2nd ed.
   1. Electronics—Charts, diagrams, etc.
   2. Signs and symbols
   I. Title
   621.3815′3′0223   TK7866

   ISBN 0-408-03010-0

Library of Congress Cataloging in Publication Data

Kampel, I. J. (Ian Joseph), 1945–
   A practical introduction to the new logic symbols.

   Includes index.
   1. Logic circuits––Notation.   I. Title.
   TK7868.L6K36   1986   621.3819′5835′0148   86-11753
   ISBN 0-408-03010-0

621·3014'8
KAM

Photoset by Butterworths Litho Preparation Department
Printed and bound in England by Anchor-Brendon Ltd., Tiptree, Essex

Author's note

Having once been a design engineer, and still being constantly in contact with design engineers, I am well aware that they will not take kindly to a new standard being imposed upon them. With the present pace of development in digital and microelectronics, there is enough to keep them busy without a radical new way of representing circuits. Unfortunately this is the price to pay for being involved in modern technology. Already the semiconductor manufacturer and military documentation use the new logic symbols. Thus there is no way of avoiding the issue. It is my hope that this book will make the transition from the 'old' symbology to the 'new' symbology as painless as possible.

There are now many national specifications being based upon the International Electrotechnical Commission publication for binary logic symbols, and by their very nature as definitive works, they cannot be as informal as a book such as this. This book aims to make the reader familiar with the new method as quickly and easily as possible, and should then serve as a handy reference source; the symbology is sufficiently complex that every engineer should need such a work at his elbow until he has had extensive experience of using the new logic symbols. Be warned: *this is far from a straightforward substitution of new shapes for old.* A totally different concept is involved.

Firstly, a word of encouragement. The news is not all bad. The new logic symbols may seem complicated at first, but they do offer designers some superb facilities for specifying designs at a high level without the need to consider precise forms of implementation; all this is achieved whilst retaining a precise specification of the required logic functions with the minimum amount of supporting documentation. I also suggest a method of working in commercial organizations which is directly associated with different levels of representation in the new logic symbology; so a design can pass down the chain from system level to final implementation in three stages of refinement. Complex circuits can be reduced to compact forms of representation which more clearly represent the logic functions performed.

The new logic symbology should prove to be a boon to digital engineers once they become familiar with the principles. Hopefully this book will help them to do just that, and will be of assistance as a general reference source thereafter.

Finally I should like to add that because of the enormous potential for discussion that the subject matter of this book provokes, it is with regret that I must state that I cannot enter into any discussion or correspondence on the subject. Readers are best advised to firstly consult their own country's standard and to then consult the body publishing that standard with reference to a particular item within that standard in order to resolve any query they may have. Alternatively, they might wish to approach the IEC directly in connection with IEC Publication 617–12 : Binary Logic Elements, at the following address: *Bureau Central de la Commission Electrotechnique Internationale, 3, rue de Varembé, Genève, Suisse.*

Ian Kampel

Acknowledgements

The author is greatly indebted to Mr D. B. J. Hicks of the British Standards Institution for his kind assistance, and to Mr C. J. Stanford (General Secretary) and Mr L. van Rooij (Deputy General Secretary) of the Central Office of the International Electrotechnical Commission (IEC) for their invaluable assistance in providing advance information on IEC Publication 617: Graphical Symbols for Diagrams, Part 12: Binary Logic Symbols, and for liaison assistance with respect to my communications with other related bodies throughout the world. I should also particularly like to thank Mr P. D. C. Reefman, Chairman of IEC/SC 3A, for his kind assistance. Thanks are also due to the General Secretariat of the European Committee for Electrotechnical Standardization (CENELEC), and to Neil Warnock-Smith of Butterworths, whose word-processed communiques to diverse countries made it possible to present information on the international implementation of this standard for logic symbols. Thanks are also due in this Second Edition to Thomas Smith of the Digital Equipment Corporation and John Waters of British Aerospace for their kind interest and advice. Lastly, but by no means least, I should like to thank Graham Owen, Cathy Gilmore and Judith Molyneaux of Intereurope Technical Services Ltd., who produced all the final illustrations

Copyright

The symbols used within this book are intended for general use and do not in themselves retain any copyright. The text describing these symbols is essentially original within this book, but is broadly based upon information contained within IEC Publication 617:12; due acknowledgement is therefore given to the IEC for their kind permission in allowing me to draw upon their general definitions as the basis for the definitions used within this book. (*See also* page iv).

Contents

PART 1 INTRODUCING THE SYMBOLS

1 An international language for digital engineers 1–1
Layout of the book 1–5

2 Definitions 2–1

3 Composition of the symbol 3–1
Combination of symbols 3–2
Embedded symbols 3–3
Common output element 3–3
Arrays 3–4
Direction of information flow 3–4

4 Inputs and outputs 4–1
Polarity 4–1
Dynamic input 4–5
Postponed output 4–6
Bi-threshold (hysteresis) input 4–7
Extension input and extender output 4–7
Non-logic and analogue connections 4–8
Internal (virtual) input and output 4–9
Fixed-mode input and fixed-state output 4–9
Special outputs 4–9
Other input/output symbols 4–11

5 Simple combinative elements 5–1
Simple gates 5–1
More complex elements 5–3
Odd/even (parity) elements 5–4
Distributed connection 5–6
Buffers with extra driving capability 5–6
Hysteresis gates 5–6
Examples 5–7

6 Delay elements 6–1
Specified delay 6–1
Tapped delay elements 6–2
Unspecified and variable delays 6–2

7   Dependency notation   7–1
I/O abbreviation   7–2
Conventions   7–2
G (AND) dependency   7–3
OR dependency between affecting I/O with identical
   labels   7–5
V (OR) dependency   7–6
N (Negate) (X-OR) dependency   7–7
Z (interconnection) dependency   7–7
S (Set) R (Reset) dependency   7–8
C (Control) dependency   7–10
EN (Enable) input and EN dependency   7–11
M (Mode) dependency   7–12
A (Address) dependency   7–14
Summary of dependency notation   7–16

8   Label sequences   8–1
Input labelling   8–1
Output labelling   8–3

9   Two-state elements   9–1
The latch   9–1
Bistable elements   9–1
Monostable elements   9–4
Astable elements   9–6
Uncommon bistable elements   9–7

10   The common control block   10–1

11   Shift registers and counters   11–1
Shift registers   11–2
Counter/dividers   11–5

12   Coders   12–1
Input and output codes   12–1
Coders   12–3
Code converters   12–4
Embedded coders   12–7

13   Signal level converters   13–1

14   Selectors   14–1
Multiplexers   14–1
Demultiplexers   14–3

15   Memory   15–1

16 Arithmetic elements   16–1
Adders and subtractors   16–1
Carry look-ahead generator, multiplier, comparator
and ALU   16–3

PART 2 USING THE SYMBOLS
17 A complex symbol analysed   17–1
Examples of complex devices   17–5
Gray boxes   17–7

18 Different levels of representation   18–1
The levels   18–1
Three levels – three uses   18–2
An example   18–3
The new logic symbols in documentation   18–10
The new logic symbols in commercial organizations   18–11

PART 3 CLOSELY RELATED SYMBOLS
19 Multiple signal paths   19–1
Signal lines   19–1
Grouped signals   19–2
Bus interconnections   19–3

20 The changer symbol   20–1

APPENDICES
  1 Glossary of symbols and notations   A1–1
Table 1 – Qualifying symbols for inputs and outputs   A1–2
Table 2 – Symbols used inside the outline   A1–3
Table 3 – General qualifying symbols   A1–7
Table 4 – Dependency notation   A1–9

  2 Participating countries   A2–1
Derivation of the IEC standard   A2–1
Degree of conformity   A2–2
Note concerning conformity   A2–4

  3 Changes in this edition   A3–1

INDEX

Part 1
Introducing the symbols

Part 1
Introduction to the spindle

1
An international language for digital engineers

Digital technology revolutionized the field of electronics: so *the new logic symbology* will revolutionize the engineer's way of depicting logic circuitry. Any revolutionary scheme is bound to have its opponents since it disrupts a happy norm, but so long as such a scheme has a sound foundation, it should soon have its equal share of adherents. It is perhaps dating to label anything 'new' these days, but this change in logic symbology is so significant that any generation of engineers who have to convert from the 'old' to the 'new' are bound to always think of the latter as the 'new logic'. At the time of revising this edition, the new logic symbology is so new that it has yet to make a real impact on the average digital engineer. It is mainly for this reason that the 'new' label is appended: to impress upon such an engineer that there really is something *new* for him to get to grips with.

Until an engineer seriously investigates the new logic symbols or hears something about them, he may be tempted to imagine that it is no more than an alternative method of drawing his particular well-beloved symbols. Such a fond (but naïve) belief may cause him to put off looking into the matter until he finds himself becoming increasingly baffled by manufacturer's data sheets which employ the new symbology. Sooner or later there will come a real shock, and he will find himself in the embarrassing position of having to temporarily suspend his current work in order to undergo a crash course in the new logic symbology. The temptation is to ignore the new logic symbology in the hope that it may go away, but this just will not happen: too many countries are introducing the common agreed standard. Each and every digital engineer is strongly urged to grasp the nettle *now* and thereby learn how the new logic symbology can help him rather than wait until the time is reached when ignorance of it will seriously hinder him.

Can you imagine the day when there is an international language? People have tried with over one hundred artificial languages, and only two of these have been particularly noteworthy: Volapük and Esperanto. Of these, the average person only recognizes Esperanto, and whilst it has its adherents from Manchester to Moscow, there is not sufficient incentive for it to catch on in a big way. Contrast this with a language which is likely to become popular in America, Britain, France, Germany, the Netherlands, Scandinavia, Norway, Austria, Belgium, Finland, Italy, Sweden, Switzerland, Canada and Australia (to mention but a few)! Such a language exists: it is the 'language' of *the new logic symbology*. It is also a language which has a real incentive for people to learn it: *in due course their career may depend upon it!*

'Language' is not really too strong a term to use, for the new symbology is a truly flexible descriptive method used to express thought processes; like a language, it is possible to express the same concept in several different ways. Similarly, it is possible to be either brief and concise or to labour a point.

It should be clearly understood from the outset that the new logic symbols are not simply one-for-one replacements of existing symbols. At their lowest – and least sophisticated – level this can be the case, but there would be little need for a book such as this if all it was necessary to do was to provide a glossary of equivalent symbols. The real value of the new symbols is in their ability to reduce a typical logic circuit into a more readily understandable form, and thereby more clearly represent the circuit functions.

As any digital engineer will appreciate, a great many logic circuits contain parallel functional elements which are identical; this is the case when performing an operation on a number of different bits, for example. Instead of showing each individual bit separately in such a case, it is possible to group the similar elements into a single symbol. It can be seen immediately that if all duplication is removed from a logic diagram, this goes a long way towards simplifying it.

Whereas before a block diagram was needed to reduce a large logic diagram into a readily understandable form, now the new logic symbols make it possible to achieve the same end *whilst retaining all the logic detail*. Block diagrams are dependent upon associated text to explain the function of the elements represented, but the new logic symbology retains a precise definition of the logic operations performed without the need for supporting text. Since this obviously defeats all language barriers, it is perhaps not surprising that so many different countries want it to work.

It may be seen from the foregoing that it is possible to employ the new logic symbology at different functional levels. At the lowest level it may represent individual gates and/or individual integrated circuits, and may contain full pin information. At an intermediate level, it may be optimized for understanding, by combining all similar functions into single symbols. At the highest possible level, it may be possible to reduce an entire complex circuit to a single complex symbol.

The level that any person cares to use really depends upon his objectives. If the aim is to show the precise method of implementation, then the level should be chosen such that all integrated circuit pin information may be shown conveniently. If the object is to explain what functions are performed by a circuit, an intermediate level is preferable, where the circuit is reduced to the most concise form possible without requiring the user to have to puzzle out the symbology. If the requirement is to produce the most concise symbol possible to express the complex functions of a complex device – as a manufacturer may wish to do on a data sheet – then the highest level of reduction may be employed.

Manufacturers are already employing highly condensed symbols to represent complex digital devices. The advantage is that the original labelled rectangle which needed supporting notes is thus replaced by a labelled symbol which fully defines the functions of the device, and that same symbol may be used on a logic diagram to completely represent the functions performed; in theory, the user should never need to go to a separate data sheet in order to work out the functions of a device so represented. The disadvantage of this approach with really complex devices is that they begin to enter the realms of the incomprehensible!

The concept of the new logic symbology allows logic symbols to be embedded inside other logic symbols; this implies (and allows for) several logic symbols to be 'nested' within each other, i.e. symbol A contains a symbol B, symbol B contains a symbol C, etc. This principle is employed by manufacturers to represent complex devices, but whereas there might be some justification for this when the aim is to represent a single device and show its pin connections, it cannot be a recommended procedure for condensing circuits formed from a number of separate components if this leads to difficulty in interpreting the symbol.

If the purpose of a logic diagram is to explain how a circuit operates then it is defeating the object to condense it into a few highly complex symbols. No engineer should be tempted to regard the creation of symbols as an academic challenge to produce the most condensed symbols possible; such clever tricks will not be appreciated by fellow engineers and certainly not by service engineers. Instead the aim should always be to optimize the symbology to most clearly represent the logic functions being performed. In the same way, the 'dependency notation' should not be employed to such a degree of complexity that the user requires a higher mathematics degree to interpret it. Engineers have a reputation for being practical; physicists would no doubt delight in making the new logic symbols more and more complex, but if the new logic symbology is to be a success, engineers must never follow that dangerous route.

The new logic symbology can be of tremendous benefit to engineers the world over. Not only does it simplify documentation methods and promote understanding, but it can greatly assist design engineers: it enables them to design at higher levels, without being device specific, and yet still produce a design which may be subsequently and unambiguously converted into a detailed design implementation matching up to any particular device preferences or constraints. All stages could be carried out by a single designer, using the method as a means of improving efficiency and of gradual refinement, or a systems or design engineer could undertake the high level design and then pass this on to other engineers to implement in terms of actual devices.

Another benefit of the new logic symbology is that it can even cope with an interface between circuits previously drawn in opposing logic conventions (i.e. positive and negative logic). This is achieved with the minimum of fuss by employing 'polarity indicators'.

The new logic symbols are intended for use in many different technical disciplines, e.g. electronic, electrical, pneumatic, hydraulic, mechanical, etc. Its primary use is probably in the field of digital electronics, and this is the only discipline covered by this book, although the principles described herein are equally applicable in other disciplines for logical situations.

It is hoped that engineers will recognize that the temporary inconvenience of having to learn new methods of representation will be amply rewarded by an understanding of a system which is capable of directly relating to the functions which they represent; previous standards did just that in the early stages, where individual gates were represented, but technology has outgrown the symbology, and anything remotely complicated has become a rectangular *black box*. The new logic symbols take a step forward to catch up with technology. The next generation of digital engineers will be taught how to use the new logic symbols. The present generation of digital engineers must get to grips with the new logic symbols without delay if they wish to survive. Just like a second language, understanding of the new logic symbols only comes with constant use and familiarity.

This book is based upon IEC Publication 617-12: 'Binary Logic Elements'. This may be regarded as the base document upon which the standards of participating countries are built. It is therefore a prime reference source and the best basis for a book aimed at a world market; in this way it avoids the risk of misleading due to the inclusion of a country's peculiarities or the omission of any essentials. The latter point is particularly relevant, for early issues of standards tend to omit anything on which they wish to ponder further; from the practical point of view, it is better to use a symbol which may be subject to minor change than be without a symbol that is urgently needed.

It is hoped that this book will therefore assist engineers, technical authors and students the world over to understand the practical application of the new logic symbols. Once they have a good understanding of the principles involved, they are advised to check with the relevant standards of their own country for any deviations in usage. It is hoped that there will be few, otherwise long years of debate within the International Electrotechnical Commission will have been in vain. Refer to Appendix 2 for a list of participating countries, and allow for the fact that some countries may take some time to fully incorporate all the principles of the new logic symbology.

The success of the new logic symbols depends entirely upon the attitude of its users. Providing that they recognize that it is a necessary step in line with advancing technology and consequently accept it, it will be a major triumph of international co-operation. It is now up to engineers: use it, don't abuse it, and it will surely work!

Layout of the book

I have likened the new logic symbology to a language; because of its involute nature it presents similar problems to teach. Such is the interdependency between different aspects that discussion of any given aspect in early chapters is limited by the reader's lack of knowledge of other related aspects. There is only one solution to this problem: early chapters need to consider certain topics in a fairly elementary form which is enlarged upon in later chapters when more general knowledge has been attained. It must therefore be appreciated that it is not possible for someone without any previous experience of the new symbology to expect to dip into later chapters and read them with full understanding; this can only be achieved when they are read consecutively.

Think of the new logic symbology as a new technical language and be prepared to start with simple phrases, only developing later into a full appreciation of the finer points of its 'grammar'.

2
Definitions

In order to avoid ambiguity it is essential to establish definitions of terms to be used within this book. All critical definitions are contained within this chapter; they should be read and fully understood before reading subsequent chapters, and should be referred to where any doubt exists in the reader's mind when reading following chapters. Rather than arrange the following definitions in alphabetic order, they are presented in a more logical order. Note that subsequent chapters amplify many of the definitions.

Logic function

A *combinative, storage, delay* or *sequential* function expressing a relationship between signal input(s) and the resultant output(s).

Logic symbol

The graphical representation of a logic function.

Pure logic diagram

A diagram that depicts logic functions primarily by means of logic symbols; all logical relationships are shown in the simplest manner, without reference to physical implementation.

Logic diagram

A diagram that depicts logic functions primarily by means of logic symbols; the details of all logical relationships, signal flow and control are all depicted, but not necessarily the point-to-point wiring.

Circuit diagram

A diagram that depicts all circuit details by means of logic and other symbols, and contains the details of all logical relationships, signal flow and control; supplementary notations are employed to fully define the physical implementation of the circuit.

Binary variable

A variable which has only two discrete values; it may be equated to any physical quantity for which two discrete levels (or ranges of value) can be defined.

Binary logic element

An element whose input and output quantities represent binary variables and whose outputs are defined digital functions of the inputs (*see Figure 2.1*).

States of a binary variable

The two values of a binary variable are assigned *logic states*.

Logic states

The two possible logic states of a binary variable may be represented by any two arbitrary symbols; the convention is to use the symbols 1 and 0 for this purpose; the terms 'true' and 'false' may be applied respectively. When external logic states are used, the *logic convention* must be defined.

Logic levels

In practice, two logic states may be represented by two different but discrete ranges of value for a particular binary variable. For example, in TTL logic, the binary variable is represented by a voltage ranging from $0\,V$ to $+5\,V$, but a logic 1-state is defined as any voltage within the range $+2.0\,V$ to $+5\,V$, and a logic 0-state is defined as any voltage within the range $0\,V$ to $+0.8\,V$ (NB. This assumes a *positive logic convention*); in such a system, a voltage within the range $+0.8\,V$ to $+2.0\,V$ is invalid. It is customary to define a nominal value for each logic level, e.g. $+3.3\,V$ for the logic 1-state and $+0.2\,V$ for the logic 0-state in a positive logic TTL system.

In order to avoid specific reference to a given method of implementation, symbols may be assigned to represent logic levels; it is customary to use the symbol H for the level with the most positive algebraic value, and the symbol L for the level with the less positive algebraic value.

The use of logic levels as opposed to logic states obviates the need to define the *logic convention*.

Logic convention

A convention which established the relationship between each external logic state and its corresponding logic level. (See next two definitions.)

Positive logic convention

In *positive logic* the H (high) level of a physical quantity (e.g. voltage) represents the logic 1-state of a binary variable, and the L (low) level represents the logic 0-state. (Negative logic is the converse.)

Negative logic convention

In *negative logic* the H (high) level of a physical quantity represents the logic 0-state of a binary variable, and the L (low) level represents the logic 1-state. (Positive logic is the converse.)

Polarity indicator convention

In the polarity indicator convention, *polarity indicators* must be employed to indicate logic levels at the input(s) and output(s) of binary logic elements. Lines *external* to a device are said to be at a *logic level* rather than a *logic state*.

Polarity indicator

A symbol used at the input(s) and output(s) of binary logic elements to show that the (external) L-level corresponds to the internal 1-state.

Combinative element

A combinative (binary logic) element is an element usually containing a *general qualifying symbol.* A combinative element has only one set of output states for any one set of input states. (This, therefore, includes AND, OR, etc.)

Storage element

A storage (binary logic) element is an element capable of storing a desired binary variable (e.g. a memory element).

Delay element

A delay (binary logic) element is an element in which specified transitions at the input give rise to one and only one related but delayed transition at the output.

Sequential element

A sequential (binary logic) element is an element for which the output(s) may be modified (sequentially) by the application of clock or control input pulses, e.g. bistable element, counter, shift register, etc. Thus a sequential element may have several possible sets of output states for any one set of input states.

Bistable element

A bistable element is a binary sequential element with two stable states.

Internal logic state

The logic state assumed to exist inside a symbol outline at an input or output (*see Figure 2.1*). It is incorrect to refer to *logic levels* within a symbol outline.

External logic state

The logic state assumed to exist on the outermost side of an external *qualifying symbol* appended to the input or output of a binary logic element (*see Figure 2.1*).

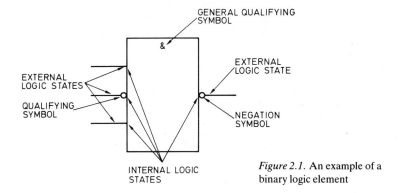

Figure 2.1. An example of a binary logic element

Qualifying symbol

Symbol used at the input or output of a binary logic element as a qualifier (*see Figures 2.1 and 3.1*).

General qualifying symbol

Symbol used within the outline of a binary logic element to define the function performed by that element (*see Figures 2.1 and 3.1*).

Negation symbol*

A negation symbol is a qualifying symbol in the form of a small circle, as depicted on the output line in *Figure 2.1*. This implies inversion of the signal. Thus in the example given, a true internal logic state gives rise to a false external logic state, and a false internal logic state gives rise to a true external logic state. An input or output without a negation symbol may be referred to as a true input/output, and an input or output with a negation symbol may be referred to as a false input/output.

* This phrase is used for clarity in this book, although 'negation symbol' is not an official IEC term.

Conclusion

The definitions given within this chapter are for terms used widely throughout the book. No attempt has been made to include the definition of terms restricted to specific types of binary logic element.

3

Composition of the symbol

Symbols are used to represent binary logic elements. This chapter describes the rules applicable for the composition of such symbols.
  A symbol comprises:

(a) An *outline*;
(b) A *general qualifying symbol* denoting the function of the binary logic element;
(c) *Input* and *output lines* with any associated *qualifying symbols*.

Figure 3.1 depicts the general case for symbol composition. The outline is rectangular; although length-width ratios are arbitrary in terms of standards, the following principles should be borne in mind in order to maintain a consistency of construction:

(a) Except where the outline is a square or the detail of content prevents this, aim to keep the width a constant measurement.
(b) Except where specifically stated to the contrary, aim to keep rectangular outlines with the longer edges vertical.

The preferred location for the *general qualifying symbol* (defined in Chapter 5) is at top dead centre of the outline, although a central location is allowed. Since there appears to be no

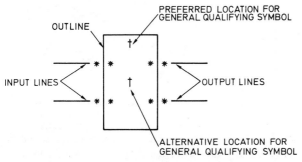

* POSSIBLE LOCATIONS FOR QUALIFYING SYMBOLS

Figure 3.1. Symbol composition

good reason why the preferred location should not always be used, this preference is applied throughout this book; it is recommended. The † symbol is used to represent a general qualifying symbol.

It is preferred to have input lines on the left of an outline and output lines on the right of an outline, so maintaining a left-to-right flow. Unless otherwise unavoidable, inputs and outputs should always be placed on opposite sides of the outline. In special cases where a horizontal symbol is approved (e.g. counters and shift registers), a top-to-bottom data flow is preferred.

It should be noted that a general qualifying symbol is normally required within an outline to specify the function of the element. In some circumstances, where the function of an element is completely determined by the qualifying symbols associated with its inputs and/or outputs, no general qualifying symbol is needed.

Combination of symbols

In order to reduce space requirements on diagrams, separate symbols for basic operations may be abutted provided that the following rules are obeyed:

(a) There is no logic connection between elements where the common side(s) of their outlines is in the direction of signal flow.

(b) There is at least one logic connection between the elements where the common side(s) of their outlines is perpendicular to the direction of signal flow.

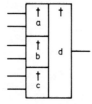

† GENERAL QUALIFYING SYMBOLS   *Figure 3.2*. Combinations of symbols

Thus in *Figure 3.2*, there is at least one logic connection between element d and each of the elements a, b, and c, although there is no logic connection between elements a and b or between elements b and c.

Embedded symbols

Another way of saving space on diagrams is to embed one symbol inside another, as illustrated in *Figure 3.3*. Any logic symbol may be placed inside another logic symbol providing that the result is unambiguous, and that the relationship between the two is clearly defined either by position or by internal connection lines.

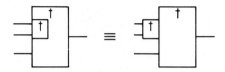

Figure 3.3. Embedded symbols

Common output element

Where a number of outputs form common inputs to a further element, this may be shown by means of a *common output element*; the basic outline for this element is shown in *Figure 3.4*.

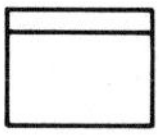

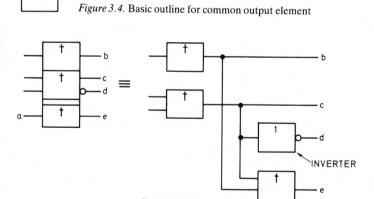

Figure 3.4. Basic outline for common output element

Figure 3.5. An example of a common output element in use

Figure 3.5 provides an example of the common output element in use; thus in this example, true outputs b and c of the associated combined elements form inputs to the common control element, which also has an independent input a. Note that it has been necessary to include a negation symbol on one output line in order to clarify the situation with regard to the false output and the actual inputs to the common output element.

Arrays

It has been described how separate symbols for basic operations may be abutted. The left-hand side of *Figure 3.6* shows that where the various elements have an *identical* general qualifying symbol, this need only be shown in the first (uppermost) element. If the abutted elements have differing general qualifying symbols, these should be shown individually to avoid any ambiguity.

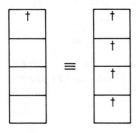

† IN THIS INSTANCE THE
GENERAL QUALIFYING
SYMBOL MUST BE IDENTICAL   *Figure 3.6.* A common array of identical elements

Direction of information flow

The preferred direction for information flow is from left to right and from top to bottom of a diagram. If this is not possible and the direction of information flow is not obvious, then lines carrying information should be marked with an arrowhead. *Figure 3.7* depicts a case of data flow from right to left, and also illustrates a case where a single line is bi-directional, i.e. data flow may be in either direction.

Figure 3.7. Examples of directional arrows

4

Inputs and outputs

Every attempt should be made to place inputs and outputs on opposite sides of a symbol, with the distinct preference for inputs on the left and outputs on the right. Qualifying symbols on inputs and outputs should be adjacent to the symbol outline.

Polarity

The logic function performed by a particular binary logic element is dependent upon the choice of logic convention, as the following example shows.

Consider a hypothetical binary logic element which behaves in accordance with the following function table:

Inputs		Output
A	B	Q
$-2\,V$	$-2\,V$	$-2\,V$
$-2\,V$	$+3\,V$	$-2\,V$
$+3\,V$	$-2\,V$	$-2\,V$
$+3\,V$	$+3\,V$	$+3\,V$

If a *positive logic convention* is applied, the *high level* is represented by the most positive voltage (i.e. $+3\,V$) and is therefore equivalent to the logic 1-state; the *low level* is represented by the least positive voltage, which in this example is actually a negative voltage (i.e. $-2\,V$), and is therefore equivalent to the logic 0-state. Substituting logic states for the voltage levels given in the preceding function table gives the following truth table:

Inputs		Output
A	B	Q
0	0	0
0	1	0
1	0	0
1	1	1

From the above truth trable is may be seen that when the positive logic convention is applied an AND function is performed.

If a *negative logic convention* is applied to the same function table, the high (most positive) level represents the logic 0-state and the low (less positive – in this case *negative*) level represents the logic 1-state. Substituting for the voltage levels in the function table now results in the following truth table:

Inputs		Output
A	B	Q
1	1	1
1	0	1
0	1	1
0	0	0

From the above truth table is may be seen that when the negative logic convention is applied an OR function is performed.

The foregoing example serves to show how vitally important it is to specify which logic convention is applicable, unless this problem is avoided by using the direct polarity convention. Users of the new logic symbology are advised to observe the following rules in order to avoid any ambiguity and at the same time produce the most generally acceptable end-product:

(1) Preferably use the direct polarity convention or positive logic convention; unless a negative supply is used for logic devices, it is preferable to use the positive logic convention to the negative logic convention.

(2) Any discrete diagrams should either:
  (a) Include a note stating the single logic convention employed;
  (b) Utilize polarity indicators when appropriate on inputs and outputs where the direct polarity convention is employed.
(3) When a set of diagrams forms part of a documentation package, providing that a preceding and highlighted textual note clearly defines the method to be employed for the *entire* documentation package, it is reasonable to omit individual diagram notes about logic convention; if the convention varies from one diagram to another within such a documentation package, Rule (2) above for discrete diagrams should be observed in order to avoid ambiguity.

NOTE: Throughout this book the positive logic convention is assumed except in diagrams which employ the direct polarity convention.

The use of polarity indicators is shown in *Figure 4.1*. Note that if no qualifying symbol is present on an input or output, the internal 1-state corresponds to the most positive external level; if the polarity indicator is present on an input or output, the internal 1-state corresponds to the least positive external level. Note also that the polarity indicator points in the direction of signal flow, hence it alone is sufficient to signify signal flow.

Thus the presence of a polarity indicator on an input or output implies that the internal 1-state corresponds to an external L-level (i.e. *low*-level). This is the situation in cases (b), (c), (e) and (f) of *Figure 4.1*. Where there is no polarity indicator, the internal 1-state corresponds to an external H-level (i.e. *high*-level). Looked at from a different point of view, an input or output with a polarity indicator is active low; without a polarity indicator it is active high. Thus a diagram utilizing the polarity indicator convention depicts an interface at the symbol outline between external logic *levels* and internal logic *states*.

Where bi-directional signal flow occurs on a connecting line (signified by bi-directional signal arrows), as shown in part (g) of the figure, any necessary polarity indicator takes the preferred signal flow direction: to the right or downward, if the latter is appropriate. If the presence of polarity indicators on linked lines clearly indicates that a given line has bi-directional signal flow, as demonstrated in part (h) of the figure, the bi-directional signal arrows may be omitted.

A single diagram should never contain a mixture of inputs and outputs utilizing negation symbols and polarity indicators. The only

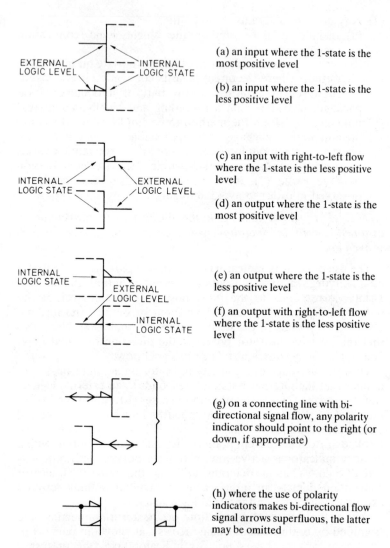

(a) an input where the 1-state is the most positive level

(b) an input where the 1-state is the less positive level

(c) an input with right-to-left flow where the 1-state is the less positive level

(d) an output where the 1-state is the most positive level

(e) an output where the 1-state is the less positive level

(f) an output with right-to-left flow where the 1-state is the less positive level

(g) on a connecting line with bi-directional signal flow, any polarity indicator should point to the right (or down, if appropriate)

(h) where the use of polarity indicators makes bi-directional flow signal arrows superfluous, the latter may be omitted

Figure 4.1. Inputs and outputs using polarity indicators

way that both symbols may appear on the same diagram is where polarity indicators are used on inputs and outputs and negation symbols are used *inside an outline. Figure 4.2* depicts legal and illegal mixtures. *Internal negation symbols apply to logic states whilst external polarity indicators apply to logic levels.*

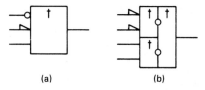

Figure 4.2. Illegal (a) and legal (b) mixtures of polarity indicators and negation symbols

The negation symbol (i.e. a circle abutting an outline of a symbol) has its conventional significance in representing an inversion of logic *state,* as illustrated in *Figure 4.3.* For a negated input, an external 1-state becomes an internal 0-state, or an

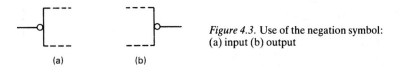

Figure 4.3. Use of the negation symbol: (a) input (b) output

external 0-state becomes an internal 1-state. Similarly for a negated output, an internal 1-state becomes an external 0-state, or an internal 0-state becomes an external 1-state.

Dynamic input

Readers will be familiar with the 'notch' qualifying symbol which indicates a *dynamic input.* The two possible states of dynamic inputs for diagrams employing negation symbols (i.e. in positive or negative logic) or polarity indicators are shown in *Figure 4.4.* The *transitory* internal logic 1-state of the dynamic input is determined by the qualifying symbol, and is true only for the duration of the specified transition of the input; at all other times the internal logic 0-state exists. Part (a) of the figure shows a dynamic input as it might be used with a single logic convention; in this case the transitory internal logic 1-state exists when the external logic state undergoes a transition from a logic 0-state to a logic 1-state. Part (b) includes logic negation on the input, therefore the transitory internal logic 1-state exists when the external state changes from a 1-state to a 0-state. Part (c) shows a dynamic input as it might be

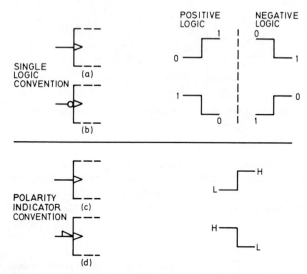

Figure 4.4. Dynamic inputs: (a) dynamic input (b) dynamic input with logic negation (c) dynamic input (d) dynamic input with polarity indicator

used with the polarity indicator convention; note that this is identical to part (a); in this case, however, the transitory internal logic 1-state exists when the external logic *level* undergoes a transition from the L-level to the H-level. Part (d) includes a polarity indicator on the input, therefore the transitory internal logic 1-state exists when the external logic level undergoes a transition from the H-level to the L-level. In all the foregoing cases the internal logic 0-state is not modified during a complementary transition.

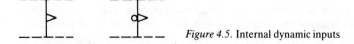

Figure 4.5. Internal dynamic inputs

An *internal* dynamic input may be shown in its two (non-inverted and inverted) forms as illustrated in *Figure 4.5*.

Postponed output

The symbol used to indicate a postponed output is shown in *Figure 4.6*. The change of state of this output is postponed until the initiating input signal has returned to its initial external logic state or level. (The reader will appreciate the significance of this symbol in conjunction with the outputs of master-slave JK flip-flops.) The internal logic state of any associated affecting or affected inputs must not change whilst the initiating input stands at its internal 1-state or the resulting output state is not specified by the symbol.

Figure 4.6. Postponed output

If the input signal which initiates the change appears as an internal connection, the change of state is postponed until the output of the preceding element returns to its initial internal logic state.

Note that both arms of the right-angle symbol should be of identical length.

Bi-threshold (hysteresis) input

The symbol used to indicate a bi-threshold input (i.e. an input with a hysteresis characteristic such as a Schmitt Trigger) is shown in *Figure 4.7.* The internal logic 0-state changes to a 1-state when the

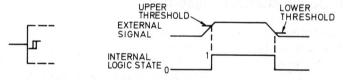

Figure 4.7. Bi-threshold (hysteresis) input

external signal level exceeds an upper threshold level, and remains at this level until the external signal level falls below a lower threshold level. A graphical representation of this is also given in the figure.

Extension input and extender output

An *extension input* or an *extender output* allows a given function to be expanded into separate functional blocks; its principal use is where actual methods of expansion are being represented, for normally, the new logic symbology would not specify such detail.

Note that the description which characterizes the relationship between external logic states and corresponding physical quantities is not normally valid for extension inputs or extender outputs.

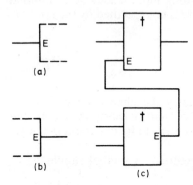

Figure 4.8. Extension input and extender output: (a) extension input (b) extender output (c) extension symbol in use

Figure 4.8 depicts: (a) an extension input, (b) an extender output, and (c) an example of the extension symbols in use. Note that in the latter case the general qualifying symbols (marked †) must be identical in both elements.

Non-logic and analogue connections

It is sometimes necessary to depict non-logic or analogue connections to a binary logic element. *Figure 4.9* illustrates how this may be achieved. Part (a) shows the symbol for a non-logic connection such as a power supply input; part (b) is an example of how a +5 V power supply input might be shown, and demonstrates than an internal qualifying label is valid. Part (c) shows the symbol for an analogue input such as the input to an analogue-to-digital converter; part (d) shows an analogue input to an element known as the 'SENSE' input.

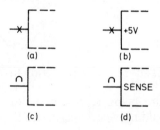

Figure 4.9. Non-logic and analogue connections: (a) non-logic connection (b) non-logic connection with qualifying internal label (c) analogue connection (d) analogue connection with qualifying label

Note that it is not normally advisable to clutter a diagram with power supply connections; these are generally more clearly represented on a separate table or diagram.

Internal (virtual) input and output

Internal or virtual inputs and outputs have no external logic state; both are depicted in *Figure 4.10*. A virtual input always stands at its internal 1-state unless otherwise affected by a dependency relationship (explained in Chapter 7). A virtual output having an

Figure 4.10. Internal (virtual) input and output

effect on a virtual input to which it is connected must be indicated by dependency notation. Note that only the dynamic qualifying symbol may be additionally appended to virtual inputs or outputs.

Fixed-mode input and fixed-state output

The symbols indicating a fixed external condition equivalent to the logic 1-state are shown in *Figure 4.11*. The fixed-mode input represents an input which must be at an internal logic 1-state for the element to perform its allocated function. The fixed-state output symbol is used to represent an output that always stands at its internal 1-state.

```
    ┌ ─ ─                      ─ ─ ┐
──┤"1"  FIXED-MODE      "1"├──  FIXED-STATE
    │     INPUT                    OUTPUT
    └ ─ ─                      ─ ─ ┘
```

Figure 4.11. Fixed-mode input and fixed-state output

Special outputs

The symbols for three-state outputs and open-circuit outputs are shown in *Figure 4.12*; it may be seen that a triangle indicates 3-state, or a diamond an open-circuit. Whilst the latter signifies an open-circuit, it does not specify the type of output; the symbols in *Figure 4.13* indicate how this may be achieved.

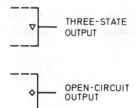

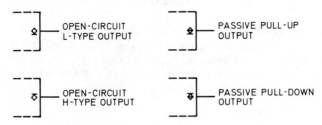

Figure 4.12. Three-state output and open-circuit output

The open-circuit L-type output produces an L-level at the output when not in its high impedance state; the H-level is achieved from the high impedance state by employing an external pull-up circuit. Conversely, the open-circuit H-type output produces an H-level at the output when not in its high impedance state; the L-level is achieved from the high impedance state by employing an external pull-down circuit. Such outputs may be wired as distributed connections (i.e. several outputs wired in parallel with a common pull-up/down circuit).

The passive pull-up output is similar to the open-circuit L-type output to its left, except that it contains an *internal* pull-up resistor. Similarly the passive pull-down output is similar to the open-circuit H-type, except that it contains an internal pull-down resistor.

Figure 4.13. A selection of open-circuit type outputs with and without internal pull-up/pull-down components (*see* text)

From the foregoing descriptions, it may be seen that the additions to the diamond symbol representing an open-circuit are as follows:

(a) An external horizontal line signifies the high/low capability of the output to go to a low impedance state;

(b) The further addition of a central horizontal line signifies an internal resistor to bias the output to the condition complementary to the low impedance state when the output would otherwise be in a high impedance state.

Other input/output symbols

This chapter is not a comprehensive list of all input and output symbols; it has concentrated upon the more common general symbols. Other chapters to follow introduce more specialized input/output symbols (e.g. for counters, shift registers, arithmetic devices, etc.). In addition, *Appendix 1* contains a summary of input/output symbols.

5
Simple combinative elements

Simple gates

Now that the preliminaries are over it is possible to consider actual binary logic elements, beginning with some simple combinative elements: these are elements in which the general qualifying symbol defines the required state of the input(s) in order to result in a particular output state(s). *Figure 5.1* is used to illustrate various combinative elements, and the references immediately following refer to sections (a) to (j) of this figure. As added assistance, type numbers have been appended in many cases since these will be familiar to the reader ('P/O' is an abbreviation for: 'part of'). Examples are taken from the 7400 series of TTL devices.

The 'AND' relationship is represented by the '&' general qualifying symbol, and in such elements, the output stands at its defined 1-state if *all* of the inputs stand at their defined 1-states. Part (a) of the figure illustrates a two-input AND gate (e.g. one-quarter of the 7408 quad 2-input AND gate device). The 'NAND' relationship is obtained in the normal way, by employing the '&' general qualifying symbol and adding the negator on the output, whereby the output stands at its defined 0-state if *all* of the inputs stand at their defined 1-states. Part (b) of the figure illustrates a three-input NAND gate (e.g. one-third of the 7410 triple 3-input NAND gate device).

The 'OR' relationship is represented by the '≥1' (equal to or greater than one) general qualifying symbol, signifying the need for *at least one* input to be at its defined 1-state for the output to be at its defined 1-state. Part (c) of the figure depicts a two-input OR gate. As before, the complementary 'NOR' gate is obtained by use of the '≥1' general qualifying symbol together with a negator on the output, as illustrated for a three-input NOR gate in part (d) of the figure. (The NOR gate requires *at least one* input to be at its defined 1-state in order for the output to be at its defined 0-state.)

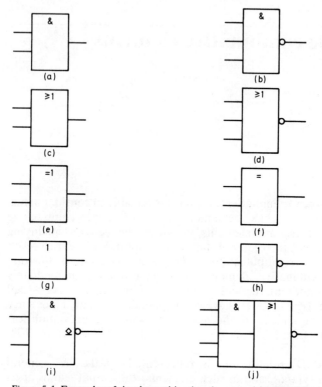

Figure 5.1. Examples of simple combinative elements: (a) 2 I/P AND (e.g. P/O 7408) (b) 3 I/P NAND (e.g. P/O 7410) (c) 2 I/P OR (e.g. P/O 7432) (d) 3 I/P NOR (e.g. P/O 7427) (e) EXCLUSIVE-OR (XOR) (e.g. P/O 7486) (f) logic identity. All inputs must be at the same state for 1-state at O/P (g) buffer non-inverting (h) inverter (e.g. P/O 7404) (i) NAND with NPN open-collector output (e.g. P/O 7403) (j) AND-OR-INVERT (e.g. P/O 7451)

The 'EXCLUSIVE-OR' (or 'XOR') relationship is represented by the '=1' general qualifying symbol, signifying that the output will stand at its defined 1-state if *only one* of the inputs stands at its defined 1-state (i.e. *one input exclusively* to be at a logic 1-state). Part (e) of the figure depicts a two-input XOR gate. The general qualifying symbol '=' is used to represent *logic identity*: the output stands at its defined 1-state providing that *all the inputs* are at either their logic 1-state or their logic 0-state, i.e. no mixture. Thus the equivalence symbol on its own implies that the input conditions are *equivalent*, be they at a 1-state or a 0-state. This is illustrated for a two-input element in part (f) of the figure; this is equivalent to an EXCLUSIVE-NOR function.

A buffer is regarded as a special case single-input OR gate, but as no ambiguity can arise with only a single input, the '≥' portion of the general qualifying symbol is omitted giving: '1', as shown in part (g) of the figure. An inverting buffer (i.e. inverter) is depicted by adding the negator to the output, as shown in part (h). (NB. A different symbol is shown for buffer *drivers* at the end of this chapter.)

Part (i) of the figure illustrates how a special output is represented by means of the appropriate qualifying symbol adjacent to the output; in this case an open-circuit L-type output is depicted – such an output is achieved with an open-collector NPN output transistor, and requires an external pull-up component in order to obtain an H-level with appropriate input conditions to the element. Part (j) shows how a two-input AND-OR-INVERT element is represented; it will be recalled from the previous chapters that there is no logic connection between elements abutted unless the common line is perpendicular to the direction of signal flow. In this example, the two input AND gates form two inputs to a following OR gate with a negated output (i.e. NOR); thus for the output to be at its logic 0-state, both inputs to either or both AND gates must be at their defined logic 1-state.

Note that in part (j) of the figure, the lower AND gate does not need to repeat the '&' general qualifying symbol; if the general qualifier is omitted in a combined arrangement such as this, it should be assumed to be identical to that in the element above.

More complex elements

The logic elements shown in *Figure 5.2* are a little more complex. The element shown in part (a) of the figure is a *logic threshold element*: the output stands at its 1-state provided that the number of inputs at their 1-state is equal to or greater than a number

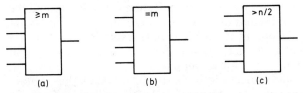

Figure 5.2. Examples of more complex combinative elements: (a) logic threshold (b) m and only m (c) majority

substituted for 'm'; thus in this example, if '2' is substituted for 'm', two or more of the four inputs must be at a 1-state for the output to be at a 1-state. Part (b) of the figure is similar; note that the 'greater than' portion of the general qualifying symbol has been removed in this case, resulting in *m and only m*. Thus the output stands at its 1-state provided that the number of inputs at their 1-state is equal in number to a number substituted for 'm'; in this latter case, substitution of '2' for 'm' requires that two and only two inputs would need to be at a 1-state for the output to be at a 1-state.

Part (c) of *Figure 5.2* depicts a *majority element*: the output stands at its 1-state provided that the *majority* of the inputs stand at their 1-states (e.g. at least three for a four input element).

When considering the more complex elements such as those depicted in *Figure 5.2*, bear in mind that the symbols are used at different levels of detail; even if there are not any specific devices available which will perform a particular logic function, a combination of devices may do so, and the simplest way of representing such a combination may be one such combinative element.

Odd/even (parity) elements

Figure 5.3 depicts odd and even elements. Part (a) shows a four-input *even element*. The output of an even element stands at its 1-state provided that an *even number* of inputs stand at their 1-state. The '2k' general qualifying symbol therefore represents an even number (since any numeral multiplied by two is even). Part (b) of the figure depicts an *odd element*, for which an *odd number* of inputs must stand at their 1-state for the output to stand at its 1-state; the general qualifying symbol '2k+1' represents odd (since 2k gives an even number, 2k+1 must give an odd number).

Part (c) of *Figure 5.3* provides an example of the even element general qualifying symbol in use for a parity generator/checker with complementary outputs (the 74280). In this example, the symbol is drawn as for the polarity indication convention: note the polarity indicator on output pin 6. If an *even* number of inputs are at an H-level, the output at pin 5 is at an H-level; the output on pin 6 is complementary to that on pin 5, therefore an L-level results on pin 6 when there is an H-level on pin 5.

Whilst on this subject, it is worthwhile noting that certain manufacturers elect to use the polarity indicator form of presentation for their devices in data sheets since this obviates the

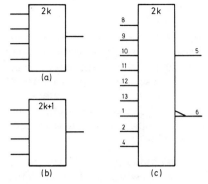

Figure 5.3. Even and odd elements: (a) even element (b) odd element (c) example of usage for parity generator/checker with complementary outputs (74280)

need to define a logic convention, and is as easy to interpret for an engineer happiest with a positive logic convention as it is for one happiest with a negative logic convention. (If you are most familiar with the positive logic convention, simply mentally transpose the negator symbol for every polarity indicator.)

It is worth noting that it is perfectly possible to represent the foregoing device as an odd element: in this case substitute '$2k+1$' for '$2k$' as the general qualifying symbol and transpose the output polarity, i.e. place the pin 5 output at the bottom of the symbol with a polarity indicator, and place the pin 6 output at the top of the figure without a polarity indicator, as shown in *Figure 5.4*.

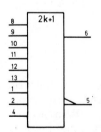

Figure 5.4. The 74280 parity generator/checker represented as an odd element

Since the outputs are complementary and pin 5 is at an H-level for an even input, clearly pin 6 is at an H-level for an odd input, hence the transposition; pin 5 is the even output and pin 6 is the odd output.

Note that the *odd element* is alternatively known by the title of: *addition modulo 2 element*.

Distributed connection

A *distributed connection* is alternatively known as a *wired function* or a *dot function*; this is the connection of outputs from a number of elements that are joined together to achieve the effect of an AND or an OR function without the use of an explicit element. The symbols utilize the diamond 'open-circuit' output symbol met in the previous chapter (*see Figure 4.13*), since wired outputs are of the open-circuit form. *Figure 5.5* depicts distributed (wired) AND and OR connections in two forms:

(a) firstly in the normal IEC recommended element form with a general qualifying symbol signifying a wired connection by the diamond;

(b) secondly in the open wired form previously in general use, with adjacent qualifying symbols (not part of the standard).

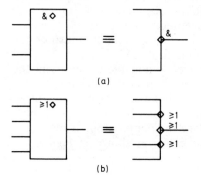

Figure 5.5. Distributed connections:
(a) distributed AND connection
(b) distributed OR connection

Buffers with extra driving capability

The triangular amplifier symbol ($\triangleright$) is used to denote the function of amplification. It may be combined with other symbols to denote outputs with extra driving capability (i.e. greater than normal fanout). See *Figure 5.7* for examples. The triangle should always point in the direction of signal flow.

Hysteresis gates

Part (a) of *Figure 5.6* shows a Schmitt-trigger NAND gate (e.g. part of 74132). The output takes on its internal 1-state only when

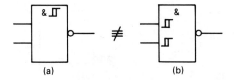

Figure 5.6. Showing the distinction between a Schmitt trigger gate (a) and a gate with separate Schmitt trigger inputs (b)

the external level applied to each input reaches its upper threshold level (*see* hysteresis inputs in Chapter 4); the output then remains at its internal 1-state until the external level applied to one of the inputs falls below its lower threshold level. (Note also that the above description refers to the *internal state* of the output, not the *external level;* the latter is inverted due to the inverter symbol on the output.) Part (b) of the figure shows a NAND gate with two Schmitt-trigger inputs; this is not quite the same thing. A Schmitt-trigger gate has a number of inputs with a logical relationship *followed* by a Schmitt-trigger; a gate with separate Schmitt-trigger inputs has independent gates with hysteresis inputs *followed* by the indicated logic function. To further this distinction, note that the 74132 two-input NAND Schmitt-trigger gate employs wire-OR diodes linked to a following Schmitt-trigger stage; a similar technique is employed for other multi-input hysteresis gates.

Examples

Further examples of specific devices are given in *Figure 5.7.*

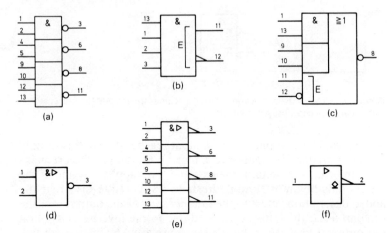

Figure 5.7. Examples of combinative devices: (a) 7400 quad 2 I/P positive NAND gates (b) ½ × 7460 dual 4 I/P expanders (c) ½ × 7450 dual AND-OR-INVERT gates, showing one gate expandable (d) ¼ × 7437 quad positive NAND buffer driver (e) 7437 quad NAND buffer driver (f) ⅙ × 7406 hex inverter buffer drivers with NPN open collector outputs

6
Delay elements

Specified delay

Delay elements are represented by an element outline containing a
horizontal line with two line terminations, as shown in *Figure 6.1*.
The left-hand line termination is labelled t_1 and the right-hand
termination t_2. Unless arrows on flow lines indicate to the
contrary, the input is always on the left and the output on the right
(although a downward flow would also be acceptable without
arrows).

The time represented by t_1 specifies the delay in response of the
output to a change at the input from a 0-state to a 1-state; the time
represented by t_2 specifies the delay in response of the output to a
change at the input from a 1-state to a 0-state. This is shown in
timing diagram form in *Figure 6.1*.

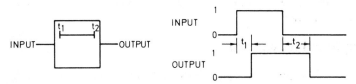

Figure 6.1. Delay element with delay times specified

The time indicators t_1 and t_2 may be substituted by actual times,
and the specification allows for these to be either inside or outside
the outline. In view of the fact that it is normal practice to place
this information inside the outline and the position of anything
outside the outline is not so easily controlled, the reader is
recommended to always insert times within the outline. Delays
may be expressed in seconds or other specified time units.

Figure 6.2 provides a selection of examples which show that it is
acceptable to specify only one of the delays. In all such instances,
unless otherwise stated, it must be assumed that *indicated delays*

are nominal. Part (a) of the figure specifies a delay of 25 ns in a 0-state to 1-state transition but does not specify the time for a 1-state to 0-state transition; part (b) does not specify the 0-state to 1-state transition but does specify a delay of 35 ns for the 1-state to 0-state transition; part (c) specifies no delay on the 0-state to 1-state transition but a 30 ns delay on the 1-state to 0-state transition; part (d) specifies a 40 ns delay on the 0-state to 1-state

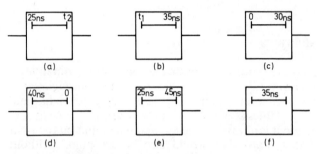

Figure 6.2. Examples of delay elements (*see* text)

transition but no delay on the 1-state to 0-state transition; part (e) specifies a 25 ns delay on the 0-state to 1-state transition and a 45 ns delay on the 1-state to 0-state transition; part (f) illustrates that if the two delays t_1 and t_2 are identical, it is sufficient to specify the time only once, centrally positioned, i.e. the example specifies a 35 ns delay in the 0-state to 1-state transition *and* in the 1-state to 0-state transition.

Tapped delay elements

Figure 6.3 facing illustrates how a tapped delay element is represented: the delay between a specific output and either polarity input transition is indicated by the time placed adjacent to the tapped outputs. An equivalent circuit is given in the figure to make this clear, where the example illustrates a tapped delay element with 10 ns tappings. (Note the three-bar equivalence sign used in this figure – it will be used extensively in following chapters to indicate *equivalent circuits* given to aid understanding.)

Unspecified and variable delays

Figure 6.2 illustrates that it is possible to use the basic symbol shown in *Figure 6.1* without specifying the delay time for both

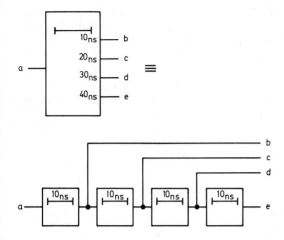

Figure 6.3. A tapped delay element

transitions, and it will be noted that since specified times are substitutes for t_1 and/or t_2, it is implied that no substitution by a specified time leaves t_1 or t_2 in the symbol. One might be tempted to ask: why not simply leave a blank where a given transition is unspecified? It is suggested that a practical reason for not doing this is to avoid the possibility of handwritten symbols with one missing delay time being confused with a delay element with equal times for t_1 and t_2, as shown in part (f) of *Figure 6.2*; clearly such a mistake could easily arise due to poor positioning, and in cramped symbols, it is a distinct possibility. Hence there is good justification for retaining t_1 or t_2 if one delay is to be left unspecified.

At certain levels of representation it may be desirable to simply represent a delay element without specifying *any* delay times. It is *suggested* that the basic symbol shown in *Figure 6.4(a)* is used for an unspecified delay; information can always be added to this at a later stage if it is subsequently decided to specify delay times.

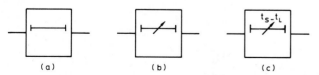

Figure 6.4. Suggested symbols for unspecified and variable delays: (a) unspecified delay (b) unspecified variable delay (c) specified variable delay

Figure 6.4 also illustrated suggested methods of representing a variable delay element by adding a slanting arrow on the delay line within the symbol to indicate variability. Although not part of the standard, the symbol shown in part (*b*) for an unspecified variable delay is internationally recognized. Part (*c*) shows a suggested form for a variable delay with specified limits for the shortest (t_S) and longest (t_L) delays: the dash is important between the two times to indicate a *range*.

7

Dependency notation

Because of the inter-relationship between different aspects of the new logic symbology, it is a great problem deciding the best order in which to explain different aspects. For example, the *dependency notation* is a means of obtaining simplified symbols for *complex elements* and should not be used to replace the symbols for combinative elements except in such circumstances; on the other hand, complex elements cannot adequately be explained without a good understanding of the dependency notation, therefore it is judged the better course to explain this now.

The dependency notation (and the refinements of labelling sequences dealt with in the following chapter), allows the relationship between inputs, outputs, or inputs and outputs, to be defined without the need for actually showing all the elements and interconnections involved. Once again it must be emphasized that it is not intended to be a shorthand (or clever) method of avoiding symbols such as those used in Chapter 5. Simple elements *should not* employ this method, and its real application will not be fully appreciated until later chapters have been read. The information provided by the dependency notation *supplements that provided by the qualifying symbols for the function of complex elements.* For this reason, examples given within this chapter depict only part of the associated complex element (i.e. broken lines signify that only part of a complex element is shown).

The terms 'affecting' and 'affected' are used in connection with inputs and outputs where dependency exists. In cases where it is not evident which inputs or outputs shall be considered as being the affecting or affected ones, any convenient choice may be made.

Due to the existence of internal feedback in some complex elements, outputs may have an effect on other outputs or on inputs.

I/O abbreviation

In order to simplify text, the abbreviation 'I/O' is substituted in the following text for the phrase: 'input or output', or for the plural form: 'inputs or outputs'.

Conventions

The dependency notation is accomplished by:

(a) Labelling the I/O *affecting* other I/O with a particular symbol denoting the relationship involved followed by an identifying number;
(b) Labelling each I/O *affected* by that affecting I/O with that same number.

Where the affected I/O already has a label denoting its function this label is prefixed by the dependency notation number.

A dependency notation number may not be used within a single element in connection with *different* dependency labels (except for A-dependency).

If an affected I/O is affected by more than one affecting I/O, then the dependency notation numbers of each of the affecting I/O shall appear in the label of the affected I/O, separated by commas. The left-to-right order of these identifying numbers shall be in the same sequence as the sequence of the affecting relationships.

The effects of affecting I/O are contained within a given binary logic element.

The numerical *values* of dependency notation numbers have no numerical significance; any symbol would serve to show a particular dependency. The convention when using dependency numbers is to start at 1 within a given binary logic element and to then utilize consecutive numbers, although any missing numbers in such a sequence has no effect on the functioning of the element.

If the labels denoting the functions of affected I/O must be numbers (e.g. the outputs of a coder, to be shown later), then the dependency notation numbers should be replaced by another character to avoid ambiguity; Greek letters are usually employed in such circumstances.

If an affecting I/O has the same dependency notation number *and letter* as another affecting I/O then they stand in an OR relationship to each other.

Dependency notation describes the relationship between internal logic states except where the EN (enable) symbol is used in connection with 3-state or open-circuit outputs; in the latter cases the enable dependency defines the relationship between the internal logic states of affecting inputs and the external conditions of affected outputs.

If the above conventions are not clear at this stage, it is hardly surprizing. Matters will clarify as the chapter progresses with numerous examples. It is important to define the conventions before explaining them more fully.

The following types of dependency are presently utilized:

(a) G – indicating an AND relationship;
(b) V – indicating an OR relationship;
(c) N – indicating a negating effect (EXCLUSIVE-OR relationship);
(d) Z – indicating an interconnection;
(e) S – indicating a setting action;
(f) R – indicating a resetting action;
(g) C – indicating control;
(h) EN – indicating an enable;
(i) M – indicating mode control;
(j) A – indicating address dependency.

Each of the above types of dependency are considered separately.

G (AND) dependency

In order to represent an AND relationship between two inputs, one must be considered as an affecting input and the other as an affected input; in this situation the choice is quite arbitrary. Thus

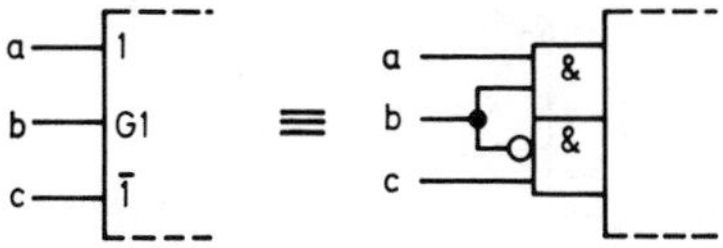

Figure 7.1. G dependency between inputs

one input into an element is given a qualifying label of G1 (the affecting input), and the other input is given the qualifying label 1 (the affected input). If it is required to show an AND relationship between an input and the complement of the affecting input, the dependency notation on the affected input has a bar imposed upon the identifying number. *Figure 7.1* depicts both cases to give inputs of a.b and c.$\bar{\text{b}}$ to a complex element. The right-hand side of the figure shows the equivalent circuit represented by the dependency notation.

The example shown in *Figure 7.2* shows that output b has an affect on input a (i.e. b is an affecting output and a is an affected input); thus input a is ANDed with output b to form an input to the complex element. *Figure 7.3* shows a similar situation where there is a negation symbol on the output; this serves to demonstrate that it is the *internal state* of the output which is ANDed with the *internal state* of the affected input.

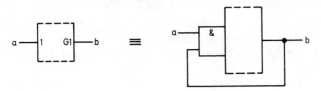

Figure 7.2. G dependency between output and input

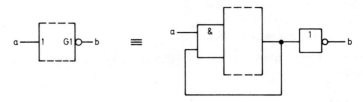

Figure 7.3. G dependency between output and input showing that it is the internal state output which is ANDed

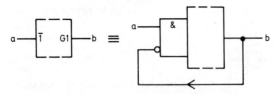

Figure 7.4. G dependency between output and input showing negation of the internal output state prior to ANDing.

An example showing an affected input ANDed with the complement of an affecting output is shown in *Figure 7.4*.

Figure 7.5 provides an example of an affecting output affecting another output; since there is a negation symbol on the b output, this also serves to demonstrate that the ANDing is an internal action prior to this external negation.

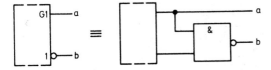

Figure 7.5. G dependency between outputs showing that ANDing is internal and prior to negation of the affected output

In the example presented in *Figure 7.6*, G dependency with a dynamic input is shown, where b is the dynamic input to a complex element (denoted by the notch qualifying symbol). Part (a) of the figure depicts the correct form of representation to show that dynamic input b only has an effect when input a is at a 1-state; part (b) of the figure is an equivalent circuit illustrating this point. It is

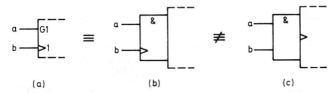

Figure 7.6. G dependency with a dynamic input: (a) and (b) are equivalent but differ from (c)

pointed out that the dependency notation does not provide the means of representing part (c) of the figure in a short-hand form, i.e. part (c) of the figure *is not equivalent* to parts (a) and (b); part (c) illustrates an AND function on ordinary inputs, the output of which forms a dynamic input to a complex element, i.e. neither inputs are themselves dynamic.

OR dependency between affecting I/O with identical labels

Before considering conventional OR relationships in the dependency notation, attention is drawn to a special type of OR dependency between any affecting I/O with identical labels, i.e. *with identical number and letter*. This is illustrated for the case where the letter happens to be G, indicating an AND dependency, but this need not be the case. *Figure 7.7* shows that inputs a and b

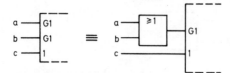

Figure 7.7. Example showing that two affecting inputs with the same letter and same identifying number stand in an OR relationship to each other (this is not restricted to G dependency and is equally applicable to affecting outputs)

have identical labels of G1; thus there is an OR relationship between inputs a and b, followed by the normal G (AND) dependency with the affected input c, as shown in the equivalent circuit.

V (OR) dependency

The V (OR) dependency operates in a similar manner to G (AND) dependency, with the letter V denoting an OR relationship between affecting I/O and affected I/O. *Figure 7.8* shows an OR dependency between the a affecting input and the internal output of a complex function giving an affected output c, as shown by the equivalent circuit; note that output b is not affected.

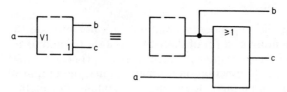

Figure 7.8. V dependency between input and output

Figure 7.9 provides a further example showing a V dependency between affecting output a from one section of a complex element and affected output b from another section of the complex element; once again, the equivalent circuit should clarify the relationship between the two outputs.

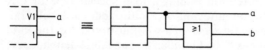

Figure 7.9. V dependency between two outputs

N (Negate) (XOR) dependency

The letter N is the symbol used to denote negate dependency. When an N dependency affecting I/O stands at its internal 1-state it complements the normal internal logic state of any affected I/O; when an N dependency affecting I/O stands at its internal 0-state it has no effect on the normal internal logic state of any affected I/O.

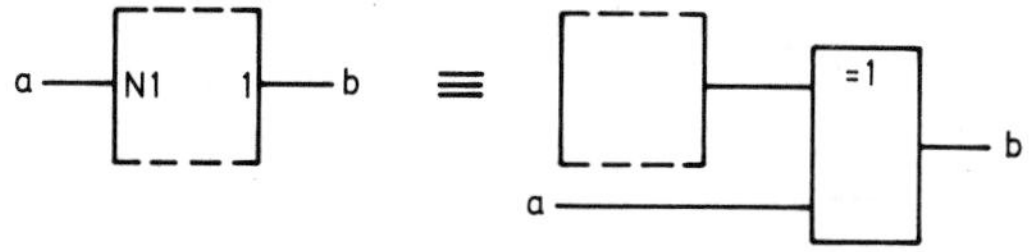

Figure 7.10. N (Negate) (XOR) dependency

It may therefore be seen that the affecting I/O stands in an EXCLUSIVE-OR (XOR) relationship to the affected I/O, as illustrated in *Figure 7.10.*

Z (interconnection) dependency

The letter Z is the symbol used to denote interconnection dependency, i.e. the existence of internal logic connections between inputs, outputs, internal inputs and/or internal outputs. The internal logic state of an affecting Z dependency I/O is imposed upon the internal state of an affected I/O regardless of its normally defined logic state. *Figure 7.11* provides three examples of Z dependency connections. The buffer symbol is used to show the transference of logic states; the partially dotted element depicts the rest of the (unshown) complex binary logic element.

Part (a) of the figure shows Z dependency between an input and an output. Part (b) shows Z dependency between the internal affecting output and a virtual input (*see also Figure 4.10*), i.e. the virtual input is the affected input in this case. Part (c) of the figure is similar to part (b), except that in this case the bar over the dependency number on the affected virtual input signifies that the affecting output has been complemented.

Figure 7.12 provides an example showing a combination of G and Z dependency; the dependency number 1 on input b is associated with G1 on input a denoting an AND relationship; the dependency number 2 on output c is associated with Z2 on input b

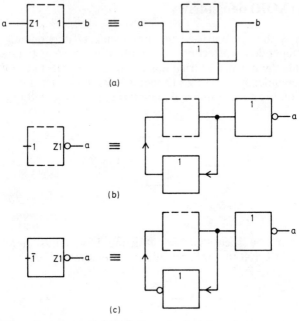

(a)

(b)

(c)

Figure 7.11. Examples of Z (interconnection) dependency

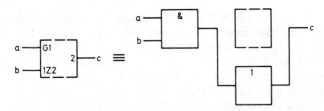

Figure 7.12. Example showing combination of G and Z dependencies

denoting that the next sequential logical event is an interconnection link to output c. Bear in mind that the inputs and outputs shown represent only some of those present in a complex element, and it is only the dependency numbers which show their association.

S (Set) R (Reset) dependency

The letters S and R are used to indicate Set and Reset dependency respectively. *Figure 7.13* provides examples of usage. Part (a) of

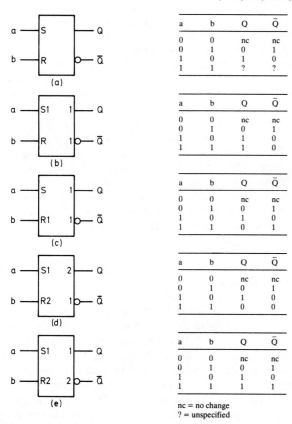

a	b	Q	Q̄
0	0	nc	nc
0	1	0	1
1	0	1	0
1	1	?	?

(a)

a	b	Q	Q̄
0	0	nc	nc
0	1	0	1
1	0	1	0
1	1	1	0

(b)

a	b	Q	Q̄
0	0	nc	nc
0	1	0	1
1	0	1	0
1	1	0	1

(c)

a	b	Q	Q̄
0	0	nc	nc
0	1	0	1
1	0	1	0
1	1	0	0

(d)

a	b	Q	Q̄
0	0	nc	nc
0	1	0	1
1	0	1	0
1	1	1	1

(e)

nc = no change
? = unspecified

Figure 7.13. S (set) and R (reset) dependencies: (a) no override (b) S overrides (c) R overrides (d) S & R override (e) S & R override

the figure shows an SR bistable (or latch) in its most basic form *without* any dependency; if S is taken to a 1-state the bistable is set (i.e. Q = 1; Q̄ = 0), if R is taken to a 1-state the bistable is reset (i.e. Q = 0; Q̄ = 1), if both inputs are in their normal condition with a 0-state there is no change from the previously specified state, but if both inputs are taken to a 1-state 'simultaneously' the result is unspecified. Such an unspecified condition does not result when dependency is introduced, since this gives rise to overriding effects.

Part (b) of the figure depicts the case where the S input overrides the R input in the event of an input clash, i.e. if both inputs are taken to a 1-state; this is specified by the S1 label on the input, and the associated dependency numbers at both outputs signify an S-override; under these circumstances the bistable behaves as it would to the combination S = 1; R = 0. Part (c) of the figure reverses the situation to give an R-override; under these circumstances the bistable behaves as it would to the combination S = 0; R = 1.

Parts (d) and (e) of the figure show that it is possible to have both S and R dependency simultaneously affecting different outputs; clearly this requires two distinct dependency numbers. In part (d) of the figure, the S input overrides the $\overline{Q}$ output and the R input overrides the Q output. In part (e) of the figure the S input overrides the Q output and the R input overrides the $\overline{Q}$ output.

It should be borne in mind that parts (d) and (e) of the figure specify the effects of S & R overrides whilst the inputs are both at a 1-state; should both inputs return to a 0-state simultaneously an unspecified complementary output condition results at the Q and $\overline{Q}$ outputs.

C (Control) dependency

The control dependency – denoted by the letter C – is used for sequential elements where more than a simple AND-relationship is implied. When an affecting C I/O stands at its internal 1-state the affected inputs have their normally defined effect on the function of the element; when an affecting I/O stands at its internal 0-state, the affected inputs have no effect on the function of that element. This is depicted for an SR bistable in *Figure 7.14* where input b is the affecting input and inputs a and c are affected inputs: the S and R inputs can only function normally when the b input is at a 1-state.

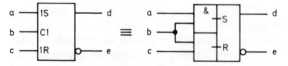

Figure 7.14. C (Control) dependency on an SR bistable

It is worth noting that in this example it would be equally valid to use the label G to denote AND relationships. Where enabling relationships are concerned with bistable elements it is normal practice to use the C label; such are *control functions* rather than miscellaneous AND relationships. This becomes clearer in later chapters when considering more complex elements.

Figure 7.15 presents a slightly more complex version of the SR bistable previously considered, and retains the same input labelling with the addition of input f, which has an AND relationship with the b input prior to becoming a control dependency affecting input. In this case the G dependency signifies a straightforward AND relationship between the f and b inputs which subsequently has a controlling dependency relationship with the a and c inputs. The sequence used within the label on the b input (i.e. 1C2) denotes the sequence that the functions take place in: since the numeral 1 is first in the label sequence it implies that G1 takes effect before C2.

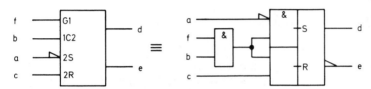

Figure 7.15. A combination of G and C dependencies

EN (Enable) input and EN dependency

The EN input is dealt with here rather than in Chapter 4 in order to contrast it with EN dependency; both are concerned with enabling. A simple EN input has an effect on *all* outputs; when an EN input is at its logic 0-state it disables (inhibits) all outputs, such that open-circuit outputs stand at their external high impedance states, all 3-state outputs stand at their normally defined internal logic states with external logic states high impedance, and all other types of output stand at their internal 0-states; when an EN input stands at its logic 1-state it enables all outputs such that all external logic states take up the condition which would exist if there was no EN relationship. This is represented in part (a) of *Figure 7.16*, where the 3-state outputs are at their normal logic state when EN is at a 1-state, or are high impedance when EN is at a 0-state.

Part (b) of the figure illustrates EN *dependency,* denoted by the fact that the EN label has a dependency notation number; thus input a (EN1) at a 1-state enables outputs d and e and also input c; input b (EN2) at a 1-state enables output f. The triangular qualifying symbol denotes a 3-state output, indicating that the outputs are in a high impedance state when not enabled by their affecting input, or their normally defined logic state (1-state or

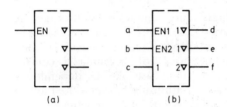

(a)   (b)

Figure 7.16. Examples of an EN input (a) and EN dependency (b)

0-state) when enabled by their affecting input. The affected input c has its normal effect on the element when enabled by affecting input a being at a 1-state, or has no effect on the element when input a is at a 0-state.

M (Mode) dependency

The useful application of M (Mode) dependencies will be seen later when they are applied to complex elements with common control blocks, yet to be explained. For this reason a 'simple' example must suffice at this stage. After a first reading of the book the reader may also consult Chapter 11 for further examples.

The effect of an affecting I/O on an affected I/O is summarized below, where separate segments of a label are separately considered if separated by solidi:

(a) if the affecting I/O is at a 1-state then affected I/O are enabled – affected inputs have their normal effect on the element (as with C dependency), and affected outputs stand at their normally defined states;

(b) if the affecting I/O is at a 0-state then affected I/O are disabled – affected inputs have no effect on the element and affected output label sets are to be ignored; if there are several sets of labels separated by solidi, any set containing the identifying number at an input has no effect on the element and should be ignored, and if at an output, those sets containing the identifying number should be ignored.

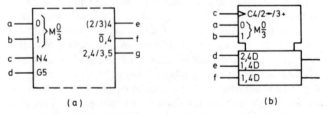

(a)   (b)

Figure 7.17. Examples of M (Mode) dependency

Figure 7.17 illustrates various points in connection with mode dependency, and also introduces a labelling method used for indicating a range of dependency numbers, i.e. $\frac{0}{3}$ in this example for the M (Mode) input. The fractional form of representation indicates modes ranging from 0 to 3 (i.e. modes 0, 1, 2 and 3); *thus, in this case, dependency numbers in this range must be reserved for mode dependency.* The bracketed section adjacent to inputs a and b has the following significance:

(a) the bit grouping symbol signifies a functional association between the two inputs a and b – in this case a decoder is represented;

(b) the 0 and 1 adjacent to the inputs signifies bit number designations for a binary number, i.e. bit 0 and bit 1; thus inputs a and b are binary inputs with input a being the least significant bit (lsb) and input b the most significant bit (msb).

Input modes for both elements are dependent upon the binary input as the following function table shows.

Inputs		Resultant mode
b	a	
0	0	0
0	1	1
1	0	2
1	1	3

For simplicity the mode dependency for affected outputs and inputs will be considered separately with respect to parts (a) and (b) of *Figure 7.17*.

Affected outputs

Referring to *Figure 7.17(a),* outputs e, f and g are affected outputs. Each is considered in turn.

Output e has a mode dependency in modes 2 and 3, and in these modes, the label set specifies negation (XOR) dependency when input c is at a 1-state due to the N4 association; in all other modes output e stands at its normally defined state as if it had no labels.

Output f also has a negation dependency when the mode dependency is operative since this is also dependent upon the N4 affecting input c; when the mode dependency is not operative the output stands at its normally defined state as if it had no labels. Mode dependency exists in the $\bar{0}$ mode, which is equivalent to specifying (1/2/3)4 in the label, i.e. mode dependency exists in all cases except Mode 0, giving an effect in Modes 1, 2 and 3.

Output g has a negation dependency with input c in Mode 2 and an AND dependency with input d in Mode 3; in all other modes the output stands at its normally defined state as if it has no labels.

Affected inputs

The circuit in *Figure 7.17(b)* has two inputs (a and b) to control which of four modes (0, 1, 2 or 3) exists at any time. The C4 label signifies dynamic control by input a over inputs d, e and f. Mode dependency indicates that input d is only enabled in Mode 2 and inputs e and f are only enabled in Mode 1. In Mode 0 nothing changes. In Mode 1, parallel loading takes place through inputs e and f. In Mode 2, serial loading through input d and shifting down occurs. In Mode 3, counting up occurs for each pulse.

A (Address) dependency

Address dependency is denoted by the letter A, and provides a means of clearly representing elements which employ address control inputs to select specific sections of a multi-dimensional array, such as memory devices. Address dependency allows the entire array to be represented when used in conjunction with a *common control block*; since the latter is yet to be explained, the general principle is applied in this chapter for a *single-bit* at a given address; it will later be shown how this can be simply expanded to apply to all bits of a given address.

In order to adequately demonstrate this principle it is also necessary to introduce a D-type bistable input, which is denoted by the letter D; bistables are considered in detail in a later chapter, but suffice it to say at this stage that the letter D denotes a D-type input.

Figure 7.18 shows a four-bit array where each single bit section is separately addressed; input e is a common dynamic clock input, f is the array input, and g is the array output. To enter data into one of the four sections of the array the following actions must take place:

(a) the appropriate A input must be at a 1-state (input a, b, c or d);

(b) the required data must be present at the array input (0-state or 1-state at input f);

(c) a clock input must be applied to input e (the 0-state to 1-state transition enters the data into the selected section of the array).

The outputs from the four sections of the array are ORed, and the A label on output G indicates that this output also has address dependency; the output from the addressed section of the array appears on output g.

Note that A in the affected input is substituted by the dependency number of the appropriate A dependency affecting input. Note also that the numerical significance of the A

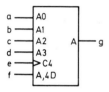

Figure 7.18. A(Address) dependency for an array with separate address input lines

dependency numbers has no other association with dependency numbers used independently of address dependency I/O and they need not therefore be unique; thus it would have been permissible to use say C1 as the qualifying symbol for input e, which would result in the qualifying symbol for input f becoming A,1D.

Because of the inefficiency of providing separate address lines for each section of an array, it is more usual to input an address to an element in binary form. *Figure 7.19* illustrates a 2-bit binary address, where a is the lsb and b is the msb; the principle previously explained for mode dependency is employed to indicate

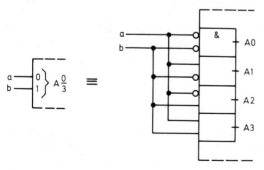

Figure 7.19. Address dependency inputs employing binary grouping

that a range of addresses is implied; the bracket indicates *binary grouping*, the A specifies an address dependency, and the 'fraction' declares the lowest over the highest address in the range. The example therefore depicts four addresses, A0, A1, A2 and A3, as indicated in the equivalent circuit.

If there are several sets of affecting inputs for the purpose of either independent or simultaneous access to sections of the array, then the letter A is modified to 1A, 2A, etc. Because they have access to the same sections of the array these sets of inputs have the same identifying numbers. *Figure 7.20* depicts two apparently very similar situations employing this principle.

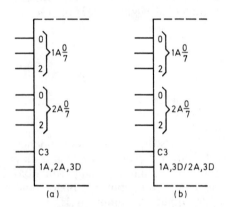

(a)   (b)

Figure 7.20. Example showing multiple set dependency (*see* text)

Both parts of *Figure 7.20* are identical apart from the data input label (i.e. the lowest input shown); the 1A and 2A binary grouped affecting inputs both address the same section of the array. The difference is that in part (a) the particular section of the array is addressed if it is selected by *both sets* of address inputs, and in part (b), the particular section of the array is selected by *either one or both sets* of address inputs. The form of labelling is explained fully in the following chapter, but briefly, the form with commas implies a sequence of logical events – thus both sets of addresses are required – and the solidi separating the two groups in part (b) implies an OR relationship, giving the one or both situation.

Summary of dependency notation

Table 4 in *Appendix 1* provides a summary of the dependency notation for quick reference purposes. The table following serves to show certain similarity of actions in the various types of dependency.

Summary of dependency notation

Type of dependency	Letter symbol	Affecting input at its 1-state	Affecting input at its 0-state
AND	G	Enables action	Imposes 0-state
OR	V	Imposes 1-state	Enables action
NEGATE (XOR)	N	Complements state	No effect
INTERCONNECTION	Z	Imposes 1-state	Imposes 0-state
SET	S	Affected output reacts as if $S = 1, R = 0$	No effect
RESET	R	Affected output reacts as if $S = 0, R = 1$	No effect
CONTROL	C	Enables action	Inhibits action
ENABLE	EN	Enables action	Inhibits action of inputs $\Diamond$ outputs off $\triangledown$ outputs high impedance Other outputs at internal 0-state
MODE	M	Enables action (mode selected)	Inhibits action (mode not selected)
ADDRESS	A	Enables action (at selected address)	Inhibits action (address not selected)

8

Label sequences

In the last figure of the previous chapter it was convenient to demonstrate the effects of label forms; this chapter further clarifies the situation and explains the significance of label sequences.

An input has a single functional effect; if a qualifying symbol is used to denote this effect is may be preceded by a label sequence. The normal reading order of a label sequence in which portions are unseparated or are separated by commas indicates the order in which the effects or modifications signified by these various portions takes effect; this has already been explained in the preceding chapter. If solidi are used to separate portions within a label, however, multiple effects are implied, and the use of this has been demonstrated in the case of mode dependency (*see Figure 7.17*). Where portions of a label are separated by solidi no significance is attached to the order of the separate portions.

Input labelling

Figure 8.1 provides examples of input labelling sequences. The left-hand portion of part (a) shows that input c has a reset action (denoted by R) when *not* in Mode 1, or has an AND relationship with input b prior to the reset action when in Mode 1; the right-hand portion of part (a) demonstrates that this is equivalent to input c having two separate mode dependency inputs to the element; both forms of representation are equally valid and the choice is immaterial. Horizontal or vertical constraints relating to symbol size might well be the deciding factor bearing in mind that portions separated by solidi do take up more than average horizontal space.

Part (b) of the figure demonstrates a convention: if one of the effects of an input is that of a normal unmodified input of the element, a solidus shall precede the first set of labels shown; this is demonstrated by the equivalent circuit.

Part (c) of the figure shows that input b has two separate effects on the element and that they are best understood by considering each separate portion of a label; the equivalent circuit in this case shows two separate effects: firstly the AND relationship between a and b, and secondly the controlling effect of input b. Remember this principle where solidi are concerned: consider each portion of the label separately, ignoring the rest.

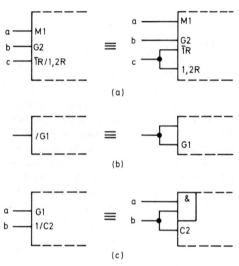

Figure 8.1. Examples of input labelling sequences

Labels may be factored using algebraic techniques, as demonstrated by *Figure 8.2*, although in such cases it is not quite so straightforward to 'ignore' portions of the label; clearly factoring must be taken into account. Part (a) of the figure shows a simple case where an input has two effects on an element: it is equivalent to two inputs, one labelled 1D and the other labelled 2D; the right-hand portion shows how 1D/2D may be factored to become (1/2)D.

Parts (b), (c) and (d) of the figure demonstrate alternative forms for an input having two separate effects: 1,2,3,4+ and 1,2,3,5−, where the notch qualifying symbol further specifies dynamic inputs; although not yet explained, the + and − (i.e. plus and minus) symbols refer to counting up and down respectively. Thus a series of actions takes place in the order shown by the dependency numbers, where actions denoted by 1, 2 and 3 take place in both cases, but when followed by action 4 this leads to incrementing of a counter, or when followed by action 5, this leads to the decrementing of a counter. Parts (b) and (c) show how separate inputs may be shown together by employing a solidus, and part (d) shows that this may be compressed by factoring.

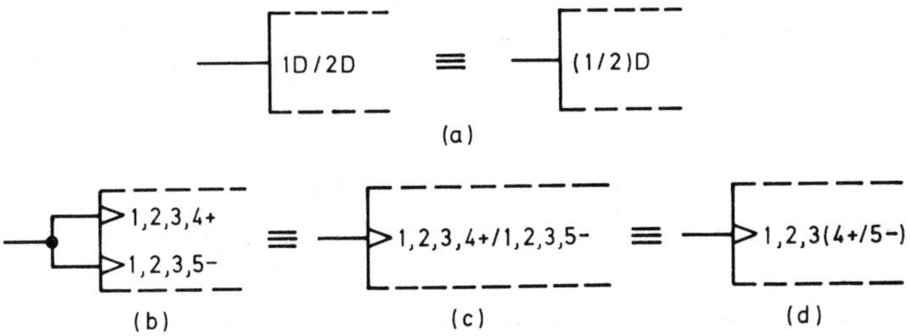

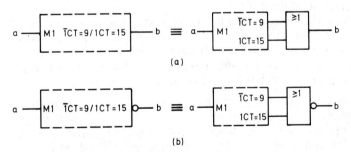

Figure 8.2. Examples of factoring input labels

Output labelling

Very similar principles apply to output labelling as to input labelling. *Figure 8.3* provides two examples. In part (a) of the figure the mode dependency input a (labelled M1) is shown to produce two separate output effects which are ORed prior to

Figure 8.3. Examples of output labelling sequences

output as b; when not in Mode 1 the CT=9 output is active, or when in Mode 1 the CT=15 output is active. The CT=m form of label denotes a decoded counter output known as a *content output*; this is one of a number of more specialized forms of output label not previously met but listed in *Table 2* of *Appendix 1*. Part (b) of the figure only differs from part (a) in that output b is negated; note that the OR effect takes place prior to the negation.

The factoring of output labels naturally follows the same rules applied for input labels, and two examples of this are given in *Figure 8.4.*

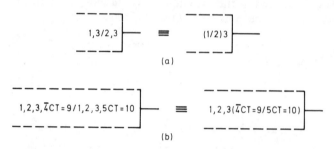

(a)

(b)

Figure 8.4. Examples of factoring output labels

A convention is applied for complex output label sequences such as the form:

$$3 \neg 1,4,5,G2$$

The convention calls for the following order of label segments from left to right:

(a) the postponed output qualifying symbol if appropriate, preceded where necessary by the indications of the inputs to which it applies;
(b) the indications of modifications to the internal logic state of the output such that the left-to-right sequence corresponds to the order in which their effects are applied;
(c) any indications that the output is an affecting output which influences associated affected inputs or outputs.

Thus in the example given above, the label specifies a postponed output with respect to input 3; modifications indicated by the numbers 1, 4 and 5 take place in that order; the output is also an affecting output having the label G2 (i.e. ANDed with other inputs or outputs as well as being a direct output).

9

Two-state elements

This chapter concentrates on two-state sequential elements and considers in turn the latch, clocked bistable elements, monostable elements and finally astable elements.

The latch

The S-R latch was introduced earlier in Chapter 7 because of the need to explain set and reset dependency notation. It is repeated here for completeness with respect to the chapter title. *Figure 9.1* depicts the basic S-R latch; in terms of familiar devices, the true or c output is the output generally defined as a Q output, whilst the false or d output is the output generally defined as $\overline{Q}$.

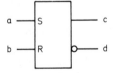

Figure 9.1. An S-R latch

Normally the S and R inputs are at a 0-state. The set or reset condition is imposed upon the latch by momentarily taking the S or R input to the 1-state respectively; it is an invalid condition to take both to the 1-state together. When the latch is set, the true output is at a 1-state; when the latch is reset, the true output is at a 0-state; the false output is complementary to the true output.

For further details of the latch and the more complicated derivatives employing overriding conditions refer to page 7–8.

Bistable elements

Bistable elements have two stable states. *Figure 9.2* depicts a number of types of bistable element and shows that the particular combination of qualifying symbols dictates the type of bistable.

The *clocked latch* shown in part (a) employs the control dependency label on the clock input which implies that the data input (not shown) only has an effect when Cm is at a 1-state; once the Cm input has returned to the 0-state the affected input can have no further effect and the bistable retains the state it was in immediately prior to the return of Cm from a 1-state to a 0-state. A specific example of such a device is given in part (c) of *Figure 9.3.*

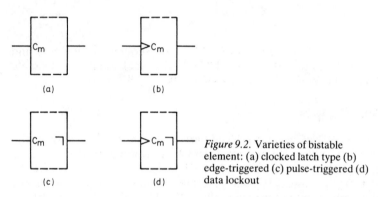

(a)   (b)

(c)   (d)

Figure 9.2. Varieties of bistable element: (a) clocked latch type (b) edge-triggered (c) pulse-triggered (d) data lockout

An *edge-triggered bistable* is represented similarly, but with the addition of the dynamic input 'notch' symbol, as shown in part (b) of *Figure 9.2.* The dynamic input (notch) implies that the associated control input only has an enabling action on the 0-state to 1-state edge; if it is desired to depict the complementary edge-triggered device (i.e. triggered by the 1-state to 0-state edge), then it is necessary to add an inverting circle to the input line. Specific examples of positive edge-triggered and negative edge-triggered bistables are shown in parts (a) and (b) of *Figure 9.3* respectively. (As always, positive logic is here assumed.)

A *pulse-triggered bistable* requires a pulse at the Cm input; this is indicated by the postponed output symbol (*see also Figure 4.6*) on the output. This combination implies that the output does not change state until the initiating input (Cm) has *returned to its initial logic state.* Since a true input is shown in part (c) of *Figure 9.2*, this implies that a negative-pulse is required (i.e. 1-state to 0-state followed by 0-state to 1-state transitions). Part (d) of *Figure 9.3* shows the converse condition for an actual device, where a positive clocking pulse is required: in this case the C1 input is initially at a 0-state but is pulsed momentarily to a 1-state; it is the trailing edge of the pulse – the 1-state to 0-state transition – which actually clocks the bistable and causes the external state of the outputs to change (always assuming that the J-K input condition calls for a change).

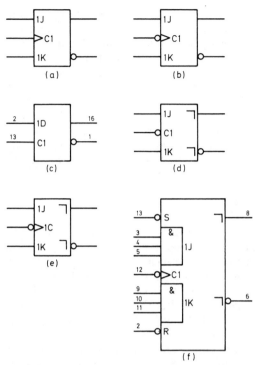

Figure 9.3. Examples of bistable elements: (a) positive edge triggered (e.g. 74109)
(b) negative edge triggered (e.g. 74112) (c) clocked latch (e.g. 7475) (d) pulse
triggered (e.g. 74107) (e) master-slave with data lockout (e.g. 74111) (f)
AND-gated master-slave with data lockout with set and reset inputs (e.g. 74110)

The problem with ordinary pulse-triggered bistable elements is
that any change at the input during the presence of the clocking
pulse may affect the final output condition. This is prevented in
the *data lockout bistable*. These devices read the data input/s at the
leading edge of the clock pulse, lock-out the data input/s for the
duration of the clock pulse thereby ignoring any change, and
transfer the read condition to the external output states on the
lagging edge of the clock pulse. Part (d) of *Figure 9.2* shows how
this is represented. A dynamic input symbol signifies that the
element is edge-triggered and the postponed output symbol
signifies that a pulse is required. In this case a negative pulse is
needed. Part (e) of *Figure 9.3* shows a master-slave bistable with
data lockout which is clocked by a positive pulse; the inverting
circle on the 1C input specifies the 1-state to 0-state transition, and
the postponed output symbols specify a pulse.

Figure 9.3 shows how the dependency notation is employed with bistable elements; the clock input becomes a control dependency input and the data input/s are affected inputs related to the control affecting input by the dependency number. Part (f) of the figure shows a more complex master-slave device with data lockout; in this case, the J and K inputs are preceded by AND gates, which are shown embedded. Set and reset inputs which are active when taken to a 0-state are also shown. Apart from these refinements, the element functions exactly as that shown in part (e) of the figure when the S and R inputs are inactive. The S and R inputs can respectively set or reset the bistable independently of the clock.

Monostable elements

A monostable element produces a single output pulse of pre-defined length. A non-retriggerable monostable produces a single pulse of fixed length in response to a triggering input. A retriggerable monostable produces a single pulse but may be retriggered during that pulse; the time-out period of a retrigger-able monostable is a fixed length after the last retriggering pulse. *Figure 9.4* shows the symbols used for the two types of monostable. A pulse symbol is the general qualifying symbol for the element, and this combined with the numeral 1 signifies a non-retriggerable monostable (i.e. only 1 triggering pulse is accepted).

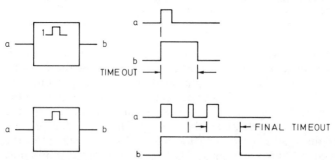

Figure 9.4. General symbols for monostable elements: (a) non-retriggerable monostable (b) retriggerable monostable

Possibly the most familiar monostable device is the 74121 (TTL device). *Figure 9.5* shows the symbol for the 74121 in part (a). Note that the general qualifying symbol denotes a non-retriggerable element. Note also that pins 10 and 11 are non-logic inputs: they are the connection points for the external timing resistor to the supply, and the timing capacitor, which is connected between these two pins; the 'X' symbol on the line indicates a non-logic connection, and the labels clarify the connection. This monostable element is triggered by either/both inputs on pins 3 or 4 being taken to a 0-state when the input on pin 5 is held at a 1-state, or by pin 5 taken to a 1-state when either/both pins 3 and 4 are at a 0-state. The embedded AND gate contains a Schmitt trigger input on pin 5, where pins 3 and 4 are not Schmitt inputs. The dynamic symbol on the output of the AND gate signifies that it is the initial AND condition which triggers the monostable element itself. When triggered, the output at pin 6 goes to a 1-state for the duration of the timeout; the output on pin 1 is always the complement of the output on pin 6 (i.e. it goes to a 0-state during timeout).

Part (b) of *Figure 9.5* shows a retriggerable monostable element, the 74123. This is seen to be very similar to the 74121. The

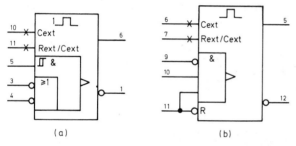

Figure 9.5. Examples of monostable elements: (a) non-retriggerable monostable (e.g. 74121) (b) retriggerable monostable with reset (e.g. P/O 74123)

principal differences are as follows: the general qualifying symbol indicates the element is retriggerable; there is no Schmitt input; the gating arrangements for the triggering input differ, and pin 11 is combined with a reset function. As before, triggering occurs dynamically when the AND conditions are first met, but now retriggering occurs if this initial condition changes but again occurs during the timeout period. For example, assume that initially pin 10 is at a 1-state and pins 9 and 11 are at a 0-state; if pin 11 is taken to a 1-state this removes the reset condition and triggers the monostable; if pin 10 is taken to a 0-state and returned to a 1-state during timeout this will retrigger the monostable; pin 11 must remain at a 1-state throughout or the monostable will be reset.

Astable elements

Astable elements produce a continuous chain of pulses at a predetermined frequency. *Figure 9.6* shows the general symbol for an astable element. The 'G' signifies a *generator*, and the two pulses signify a generator of pulses.

Figure 9.6. General symbol for astable element

An astable element may be free-running, in which case it never ceases in the production of output pulses, and the symbol is that shown in *Figure 9.6*, i.e. that of an *uncontrolled* astable element. If it is required to control the production of output pulses, a *controlled astable element* is required, as shown in *Figure 9.7*. The equivalent circuit shows that this is equivalent to an uncontrolled astable element followed by an AND gate to enable the final output; when input a is at a 1-state then output pulses are produced at the b output; when input a is at a 0-state, output b remains at a 0-state.

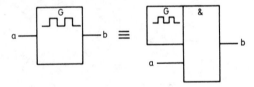

Figure 9.7. A controlled astable element

The next refinement to be considered is the relationship between the output pulses from the element and the controlling input. As *Figure 9.8* shows, this gives three options:

(a) synchronously starting;
(b) stopping after completing the last pulse;
(c) synchronously starting and then stopping after completing the last pulse.

Part (a) of *Figure 9.8* depicts a *synchronously starting* astable element. The exclamation mark inserted before the 'G' in the general qualifying symbol specifies synchronous starting. As the timing diagram shows, this means that the output pulses start synchronously with the control input being taken to a 1-state, but it does mean that the output pulse may be abruptly terminated when the control input is returned to the 0-state.

Part (b) of the figure shows how this situation is modified by an astable element required to complete the last pulse; in this case the exclamation mark is placed after the 'G'. The timing diagram shows that the final output pulse is completed even when the controlling input is terminated during the pulse. In this case the input is not synchronous, and the free-running astable element might be at any point in its cycle when input a is taken to a 1-state; consequently, the input pulse is of unpredictable length.

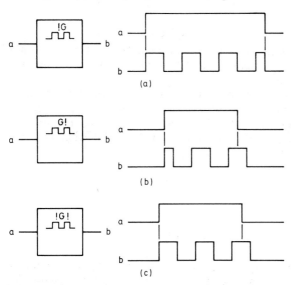

Figure 9.8. Examples of various controlled astable elements: (a) synchronously starting (b) stopping after completion of last pulse (c) synchronously starting then stopping after completion of last pulse

Part (c) of the figure puts the two features together, and is consequently represented by an exclamation mark before and after the 'G' in the general qualifying symbol. The timing diagram shows that the output starts synchronously with the control input going to a 1-state, and the final pulse is completed after the control input is returned to a 0-state.

Uncommon bistable elements

In conclusion, some uncommon bistable elements are considered. *Figure 9.9* shows SR bistable elements which have a pre-defined condition upon power-up. The I=0 general qualifying symbol indicates that the element will be in the 0-state on power-up (part (a) of the figure); the I=1 symbol indicates that the element will be in the 1-state on power-up (part (b) of the figure).

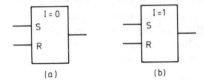

Figure 9.9. SR bistable with initial power-up state

Although not part of the specification, it is *suggested* that these same general qualifying symbols are equally applicable to other forms of bistable element if the user wishes to show a pre-defined power-up condition.

Another unusual device is shown in *Figure 9.10*; a non-volatile SR bistable, signified by NV as the general qualifying symbol. Such an element retains its logical state when power is removed; the element will therefore be in the same condition when power is switched on as it was before power was switched off.

Figure 9.10. Non-volatile SR bistable

Once again, the same argument applies for any kind of element with this property, although not specially allowed for in the specification.

The final bistable to be considered is not unusual, it is just infrequently used. The 'T' label is used to indicate a bistable 'toggle' input, i.e. an input which causes a bistable to be complemented.

Figure 9.11. Toggle input flip-flop

10

The common control block

The *common control block* is a most powerful aspect of the new logic symbology, and with the foundations which have been set by the preceding chapters, its uses will be fully appreciated after a brief explanation. It is worth noting at this point, however, that the common control block is the *only* distinctively shaped outline used in the IEC standard.

The common control block is used where a circuit has one or more inputs or outputs which are common to more than one element of the circuit. This is best understood by considering various examples.

Figure 10.1 provides a simple example of the common control block in use. Input a is to the common control block, therefore this input is common to all the elements beneath the common control block, as the equivalent circuit shows. In this example, a is an input to three separate AND gates.

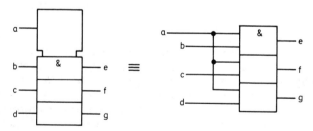

Figure 10.1. Example of the common control block in use

Remember that it is not necessary to repeat the AND general qualifying symbol in all the elements: where an array is concerned, if no qualifying symbol is shown in a particular element, it is assumed to be identical to the element above; thus it is only necessary to define the uppermost element in an identical array.

An input to a common control block without an affecting dependency notation label is common to *all* elements in the array;

a dependency notation label signifies it only affects appropriately labelled affected inputs or outputs in the array.

An example of an octal flip-flop with enable (the 74LS377) is shown in *Figure 10.2*. This comprises eight D-type flip-flops with a common clock input on pin 11, and a common enable on pin 1. Providing that the enable is low, a transition from a low-level to a high-level on the clock input enters data presented at the eight D-type inputs into the separate flip-flops. In effect, this element is an 8-bit register. The clock input is shown as a control affecting input and the eight D-type inputs are control dependency affected inputs. It is only necessary to define the input of the uppermost bistable element with the label 1D; it is assumed that the seven inputs below it are identical.

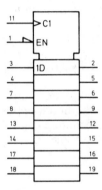

Figure 10.2. An octal flip-flop with enable (74LS377)

Note that this example uses polarity indicators. From now on some examples will use this form of representation to make the reader equally familiar with both forms of presentation. When not used, a positive logic convention may be assumed, as always *in this book*. No further comment will be made about the particular form of presentation, but the reader may care to note that where '0-state and 1-state' terminology is used for the positive logic convention examples, this is replaced by 'high/low (H/L) level' terminology when polarity indicators are used.

A 4-bit bistable latch is shown in *Figure 10.3* (the 7475). This device is arranged in two segments where the upper two elements are controlled by pin 13 (the C1 input) and the lower two elements are controlled by pin 4 (the C2 input). Compare this with part (c) in *Figure 9.3*, which shows one element of the same device. Since the input labels do differ in this array, they have been placed on all

the inputs, but it would be acceptable to delete the labels
associated with the inputs on pins 3 and 7 since each is identical to
the label above.

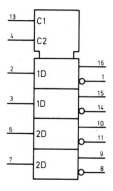

Figure 10.3. A 4-bit bistable latch (7475)

Figure 10.4 depicts a quint OR gate array with a common input
(pin 22) to each element; this device has complementary outputs.
Figure 10.5 is an example showing octal buffers/line drivers/line
receivers with non-inverted 3-state outputs (the 74S241). This
device has two separate sections, one with a non-inverted 3-state
enable and one with an inverted 3-state enable. The upper four
elements have Schmitt trigger inputs and the 3-state outputs are
enabled by an external 1-state input on the Schmitt trigger enable
input at pin 19 (EN1); the lower four elements have Schmitt
trigger inputs and the 3-state outputs are enabled by an external
0-state input on the Schmitt trigger enable input at pin 1 (EN2).

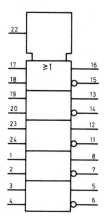

Figure 10.4. Quint OR with one common input and
complementary outputs

Note the triangular general qualifying symbol indicating a *buffer* (i.e. higher than usual driving capability).

The triangular buffer (or amplification) symbol should always point in the direction of signal flow. In *Figure 10.6* it may be seen to be reversed in the upper element, thereby signifying that signal flow in this element is from right to left. This device is a quad bus driver and is intended to also receive data from the bus; a four bit bus is connected to pins 3, 6, 10 and 13. Since data may be output to the bus or read from the bus, the device is bi-directional.

In order for the device to work in either mode, pin 1 must be in the 0-state, giving EN2 true; for the rest of this description, it may be assumed that this is the case. The state on pin 15 then dictates whether the device outputs data onto the bus or reads data from the bus. If pin 15 is at a 0-state, EN3 is true, the correct conditions occur for the dependency labels in the top element, and data is

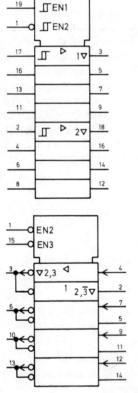

Figure 10.5. Octal buffers/line drivers/line receivers with non-inverted 3-state outputs

Figure 10.6. Quad bus driver, parallel bi-directional

therefore output onto the bus via the buffer drivers from pins 4, 7, 9 and 12; the condition 2,$\overline{3}$ set by the receiving buffer elements is false, therefore the outputs on pins 2, 5, 11 and 14 are high impedance. Thus the device cannot read the data it puts onto the bus. If pin 15 is at a 1-state, EN3 is false, and this time the condition set by the label 2,$\overline{3}$ is true, and the receiving buffers output data from the bus on pins 2, 5, 11 and 14; under these conditions the label 2,3 is false, therefore the outputs from the device onto the bus are high impedance.

Figure 10.7 depicts a dual line receiver with complementary inputs grouped together by the line grouping symbol (*see also* Table 2 of *Appendix 1*). In this example there is a common AND relationship between the input on pin 6 and both outputs, plus a separate AND input for each (pins 5 and 8). Note that it has not been necessary to repeat information in the lower element even though a different dependency number is actually required for pins 8 and 9 if such labelling is undertaken (e.g. G3 for pin 8 and 1,3 for pin 9.

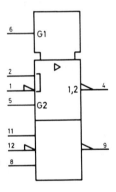

Figure 10.7. Dual line receiver

The final example is provided to show a common output; because it is common to both elements it is embedded within the common control block. *Figure 10.8* depicts a dual Schmitt trigger (inputs b and c) with true and false outputs (e and g for upper element and h and k for lower element respectively), plus 3-state optional false outputs f and j, both of which are enabled by input a in the common control block.

The common output at d is an exclusive-OR output of two outputs, and this is shown by the interconnection (Z) dependency affecting output labels on outputs e (Z2) and h (Z3) and the virtual

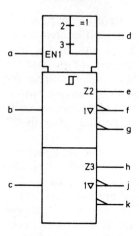

Figure 10.8. Dual Schmitt trigger with complementary outputs, inverted 3-state outputs, and a common output

inputs (labelled 2 and 3) on the common exclusive-OR gate; thus d is at a high level provided that inputs b and c are at different levels (i.e. one high and the other low). This might alternatively have been represented by employing the common output element technique (*see page 3-3*).

It may now be seen why it would be difficult to do the common control block justice without an understanding of the dependency notation.

11
Shift registers and counters

This chapter concentrates on two special types of sequential element: *shift registers* and *counters*. Each will be dealt with separately, but firstly some general points can be made about symbol conventions.

Parts (a) and (b) of *Figure 11.1* show typical symbol outlines. It will be seen that the temporary symbols α and β have been used in place of the general qualifying symbol: this is to enable the accompanying table to illustrate the correct substitutions for shift register, counter or the counter/divider.

The general qualifying symbol for a shift register is 'SRG' followed by a number giving the number of stages; thus a 4-bit shift register is labelled SRG4.

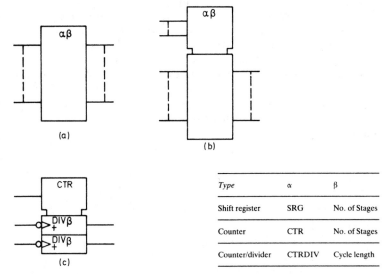

Type	α	β
Shift register	SRG	No. of Stages
Counter	CTR	No. of Stages
Counter/divider	CTRDIV	Cycle length

Figure 11.1. Symbol conventions for shift registers and counters – general qualifying symbol

The general qualifying symbol for a counter is 'CTR' followed by a number giving the number of stages; thus a 4-bit binary counter is labelled CTR4.

A counter/divider is a more complicated device: it is a binary counter with a modified count for particular division requirements (e.g. for division by 5 or by 10). The general qualifying symbol for such a device is 'CTRDIV' followed by a number indicating the cycle length; thus a *decade counter* is really a counter/divider with a cycle length of 10, and should be labelled CTRDIV10.

A complex counter/divider device may have separate elements with different cycle lengths. In this case the convention is shown in part (c) of *Figure 11.1*. The common control block retains the CTR portion of the general qualifying symbol (since counting is the common feature), and the separate elements have their own general qualifying symbol for the divider, each specifying the relevant cycle length.

Shift registers

Now to consider some further conventions regarding shift registers with reference to *Figure 11.2*, but firstly a note concerning symbol orientation. All symbols shown in this book use the preferred vertical orientation shown in this figure. Because it has been common practice in the past to use a horizontal orientation where better suited to diagrams, this practice is used to some extent and is covered by some related standards. Only vertical orientation presently features in IEC Publication 617–12, however, and since horizontal orientation can lead to some confusion regarding the direction of shift and the placement of labels, readers are advised to avoid horizontal orientation completely. Opinions – and national standards – vary!

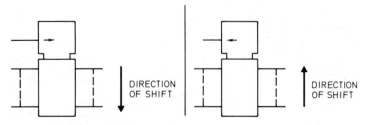

Figure 11.2. Shift register symbol conventions: (left) shift left-to-right or top-to-bottom (right) shift right-to-left or bottom-to-top

The figure is centrally divided into left- and right-hand portions; let us firstly consider the left-hand side. The internal arrow qualifying symbol indicates a shift register controlling input, and the direction the arrow points in indicates the direction of shift; thus a right pointing arrow indicates a right-shift – in the vertical symbol this equates to a top-to-bottom shift. This is easily remembered if you firstly visualize the horizontal symbol with its direction of shift and then mentally rotate the symbol through 90° in a clockwise direction.

The right-hand portion of *Figure 11.2* shows the opposite conditions which apply for a left-shift. The same principles apply.

In order to fully understand these principles it is best to study some examples, and *Figure 11.3* illustrates five shift register elements of varying complexity.

Part (a) of the figure shows a simple 8-bit shift register with serial input and complementary serial outputs; this is part of the 7491 device. Pin 9 provides the shift right control input, shown here as a control affecting input C1, which operates upon the output of the AND of data inputs on pins 11 and 12; it takes eight clock pulses on the shift input to move a given data input to the outputs.

Part (b) of the figure shows a 512-bit (static) shift register with an 'end-around-shift' mode; it serves as an excellent illustration of the use of mode control. Pin 1 is the mode control input, which initially we shall assume to be at a low level, giving M1 false; this condition enables the data input on pin 2 (since $\overline{1}$ is true), and a high-to-low transition on the control shift input of pin 5 causes a right shift (this makes the $\overline{1}$,2D label on the input of pin 2 valid), where data is entered via pin 2 and is shifted out 512 stages later on pin 3. If the mode input is at a high level, M1 is true, the serial input on pin 2 is disabled, and the virtual input labelled 3,1,2D is now true; the 3 is an interconnection dependency which feeds the output back to the input, thereby achieving a circulatory shift; thus in this mode the shift register enters data from its output. The 2D portion of the label appears in both inputs since both are dependent upon the clocking shift pulse at pin 5.

Part (c) of the figure is an 8-bit shift register with a parallel (or 'broadside') load facility (the 74165 device). Parallel loading is achieved by taking the input on pin 1 low; this is shown by the C3 affecting input label and the 3D data inputs to each element. For shifting to occur, C3 must be false, therefore the input on pin 1 must be high (this is the SHIFT/LOAD input); a high on pin 1 makes G1 true, and this combined with either pin 15 or pin 2 high establishes the conditions for shifting; when this condition first

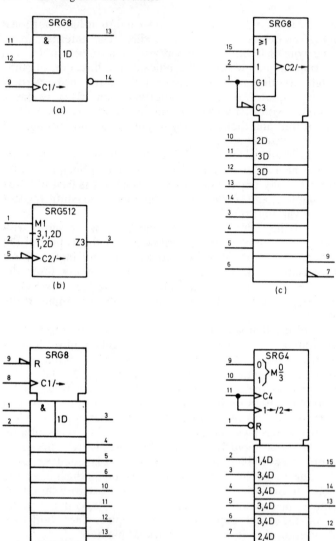

Figure 11.3. Examples of shift register elements: (a) 8-bit shift register with serial input and complementary serial outputs (part of 7491) (b) 512-bit shift register with 'end-around-shift' mode (MM4057) (c) 8-bit shift register with parallel load facility (74165) (d) 8-bit shift register with parallel outputs drawn to show method of horizontal representation (74164) (e) 4-bit bi-directional universal shift register (74194)

occurs the dynamic C2/→ condition is satisfied to give a right shift (top-to-bottom in this case). It will be seen that complementary outputs are provided on the last stage.

Part (d) of the figure depicts an 8-bit shift register with a parallel output (i.e. an output from every stage). There is little about the device to warrant special comment except to draw attention to the common reset input (R) on pin 9, the AND gate serial input arrangement, and to again point out the principle of not needing to repeat labels on *lower* elements.

Part (e) of the figure depicts a 4-bit bidirectional universal shift register (the 74194); it is universal because it can shift right or left, and it has parallel-in and parallel-out facilities. Because of this complexity it needs a more complicated mode control: four different states in all, controlled by the inputs on pins 9 and 10. Mode 0 is the inhibit clock (or 'do nothing') mode. Mode 1 is the shift-right mode, during which serial input data is entered via pin 2. Mode 2 is the shift-left mode, during which input data is entered via pin 7. Mode 3 is the parallel load mode, whereby data is loaded via pins 3, 4, 5 and 6. All clocking actions are dependent upon the dynamic C4 input. A common reset is provided at pin 1. Study this particular device to satisfy yourself that you fully understand the different actions in all its modes.

Counter/dividers

Since a counter can equally well be used as a divider, terminology can become a little confusing. For this reason a convention will be applied: the term 'counter' will be reserved for a device having a natural cycle length (i.e. unmodified to produce special counts which are not multiples of 2); the term 'counter/divider' will be reserved for devices with a modified cycle length aimed at special dividing applications.

The counting code may be shown in any convenient way such that it does not interfere with other notations, although binary code is assumed unless there is a contrary indication. It is preferred to have the least significant bit closest to the common control block, and this is assumed if not shown.

Figure 11.4 provides various examples of counters and counter/dividers which will be considered in turn.

Part (a) of the figure depicts a 14-state binary counter (CD4020) of the ripple-clock variety; this is an asynchronous device in which

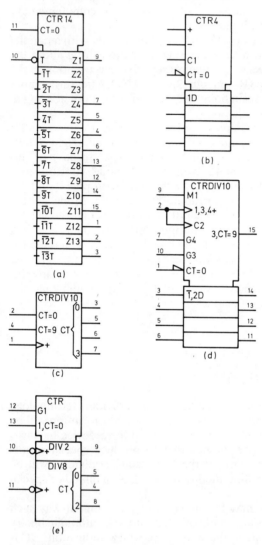

Figure 11.4. Examples of counters and counter/dividers: (a) 14-state binary counter (CD4020) (b) 4-stage bi-directional counter with parallel loading and common reset (c) decade counter (part of 74490) (d) synchronous decade counter with parallel loading, common reset and ripple carry output (e) counter/divider with divide by two and divide by eight elements (74293)

each element is toggled by the preceding element in the chain; this is shown by means of Z dependency associated with toggle (T) inputs. Only the first toggle input is externally available (on pin 10) and all the other inputs are therefore shown as virtual inputs. The device features a common reset input on pin 11, which might have been shown with an 'R' label; the CT=0 label is preferred for a counter, since this is just one special case of setting the counter; other set-count inputs may be available, in which case a CT=m label is used where m represents the count that is set. (A future symbol using the general qualifying symbol 'RCTR14' to indicate 'ripple' greatly simplifies this representation).

Part (b) shows a 4-stage bidirectional counter with parallel loading and a common reset (CT=0). Parallel loading is performed when C1 is high; normally C1 should be low. The device counts up when the + input is taken high, or down when the − input is taken high.

Part (c) shows a counter/divider with a cycle length of 10; it is normally referred to as a 'decade counter'. This counter counts up for each clocking pulse applied to pin 1; the CT=0 input may be used to reset the counter (or looking at it another way, to set it to zero); the CT=9 input may be used to set the counter to 9. The bit grouping symbol (refer to *Table 2* in *Appendix 1*) is used to indicate that the outputs are used in combination to provide the (binary) output.

Part (d) shows a synchronous decade counter with the following facilities:

(a) parallel loading clocked under control of C2 when M1 is false;
(b) common reset by taking pin 1 low (such that CT=0 true);
(c) ripple carry output when CT=9 providing that G3 is true;
(d) count-up clocked via pin 2 provided that M1, G3 and G4 are all true.

Part (e) is a counter/divider where there are separate and different dividing elements: the upper element divides by 2 and the lower element divides by 8. Each of the two elements has a separate clock input, but there is a common CT=0 input which is an effective AND of the inputs on pins 12 and 13.

Finally a special point of interest. The standard allows for a number following the + or − counter symbols to indicate a count incremented/decremented by greater than one for a single clock input. In the (unlikely) event of the reader needing to show such an effect this is the best method. No further use should be made of the same number in any dependency notation for the same element in order to avoid confusion.

Notes

This page is reserved for reader's notes on horizontal symbols should these be employed. It is advisable also to make note of the relevant standard or authority.

12
Coders

This chapter concentrates on *coders* and the special class of coder known as a *code converter*. A coder is simply a device which receives a given input code on a number of inputs and causes a related but different code to appear on a number of outputs; where there is a readily describable relationship (e.g. binary to octal), it is usual to refer to them as code converters.

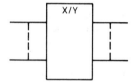

Figure 12.1. General symbol for a coder

The general symbol for a coder is shown in *Figure 12.1*; X corresponds to the input code and Y to the output code. For any given state of the coder there is an *internal number*. Depending on the input code the internal logic states of the inputs determines the internal number. The internal number is represented at the outputs by the internal logic states of the outputs depending on the output code. Such elements are therefore *combinative*.

Input and output codes

The relationship between the internal logic states of the input and the input code, or between the output and the output code is achieved by means of suitable input or output labels, or by substitution for the X or Y as appropriate in the general qualifying symbol.

Inputs are represented by numbers such that the sum of these numbers gives the internal number for a given input condition; the usual procedure is to employ binary weightings on the input lines, as shown in *Figure 12.2*. Thus a logic 1-state on inputs b and c and a logic 0-state on input a produces an internal number of 6 in this example.

Inputs			Outputs			
c	b	a	d	e	f	g
0	0	0	0	0	0	0
0	0	1	1	0	1	0
0	1	0	1	1	0	0
0	1	1	0	1	0	0
1	0	0	0	1	0	0
1	0	1	0	1	1	0
1	1	0	0	0	0	0
1	1	1	0	0	0	1

Symbol:
```
         X/Y
          1/2 ── d
a ──1
          2...5 ── e
b ──2
          1/5 ── f
c ──4
          7  ── g
```

Figure 12.2. Example of a coder. (The truth table is not necessary and is provided for illustrative purposes only)

Outputs may be represented in one of several different ways as follows:

(a) by a single number or by a list of numbers separated by solidi, where such number(s) represent internal number(s) which produce an output (*see* outputs in *Figure 12.2* for examples);

(b) by two numbers separated by three dots, where the output is true for a range of internal numbers, where the first number represents the lower limit and the second number represents the upper limit (e.g. 2...5 represents 2, 3, 4, or 5, as shown in *Figure 12.2*).

(c) by the bit grouping symbol (*see Table 2* of *Appendix 1*) and bit numbers where the Y of the general qualifying symbol indicates a standard code.

(Another method which has been used in the past (but not recommended in the IEC standard) is to label outputs with an I/O truth table, where bit grouping is used at the input to indicate lsb and msb; thus three input lines would require three columns in the truth table, or four inputs would require four columns in the truth table. No example is given in this chapter and the method is not recommended as it can give rise to confusion.)

Figure 12.2 provides an example of the most common form of input and output labelling; it is a simple and straightforward method which can cope with standard or non-standard codes. A truth table is provided in this figure purely to demonstrate the efficiency of the symbol in representing a completely non-standard code. The inputs (X) are labelled with a binary-weighted code; add the input numbers for any input to obtain the internal number, then look for the value of this internal number in the output labels to find any outputs which will respond to this internal number. Thus the combination a=1, b=0, c=1 gives the internal number of 5, and it can be seen that both the e and the f outputs respond to this particular internal number.

Coders

The example given in *Figure 12.3* is of an X/Y coder; a single output goes active for a given binary input code. As usual, add the numbers on the true inputs to obtain the internal number; the output with the corresponding number is the only output which is true.

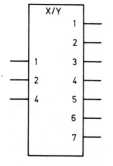

T1

	Inputs		Outputs		
1	2	3	10	11	12
0	0	0	1	1	1
0	0	1	0	0	0
0	1	0	0	1	0
0	1	1	1	0	0
1	0	0	0	1	0
1	0	1	0	0	1
1	1	0	0	0	0
1	1	1	1	0	0

Figure 12.3. A coder

Figure 12.4 A coder employing a separate table to define the internal function

There may be instances where there is just insufficient space to clearly represent the function of a coder by internal labels. In such a case an associated table must be used, and a table number reference should be placed within the symbol in square brackets (as a note). *Figure 12.4* illustrates this principle, where the note

'T1' refers the user to the accompanying table. Where such a reference is used it must be accompanied by an identically referenced table.

Code converters

Code converters are still adequately described by the term 'coders'; this particular variety of coder simply has a recognized role in the conversion between two accepted codes.

Figure 12.5 provides a simple example of a code converter which converts between a decimal input and a BCD (binary coded decimal) output. Since the input is decimal, only one input will be

Figure 12.5. A decimal to BCD code converter

true at any given time. The appropriate input gives rise to an internal number indicated by the input labels; the corresponding BCD output code gives rise to true outputs on those lines corresponding to labels with numbers needed to total the internal number. To take an example, if the input with input label 7 is the only input at a 1-state, then this gives rise to an internal number of 7; in order to match this with the output label number total the outputs labelled 4, 2 and 1 must be at a 1-state, and the output labelled 8 will be at a 0-state (as will all unmentioned inputs).

Some more complex examples of specific code converter devices are given in *Figure 12.6*. Each will be considered separately.

Part (a) of the figure depicts a Gray to decimal code converter. (Gray code is a code used for shaft encoding; in order to prevent change-over glitch problems only one bit of the input code changes

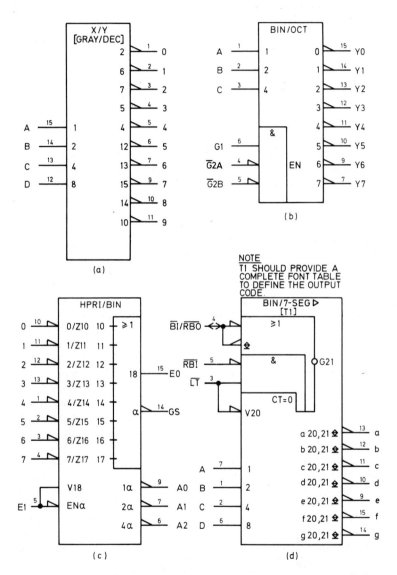

Figure 12.6. Examples of code converters: (a) 4-line to 10-line Gray to decimal converter (7444) (b) binary to octal converter (74138) (c) 8-line to 3-line priority encoder (74148) (d) BCD to 7-segment decoder/driver (7447)

when the shaft moves from one coded sector to the next.) The note in square brackets indicates the input and output codes, and the internal labelling indicates the relationship between input and output codes. Thus if there are only high-level inputs on pins 12 and 14, there will only be a low-level at the output on pin 11; the internal number applicable in this example is 10.

Part (b) of the figure depicts a binary to octal code converter, indicated by the BIN/OCT general qualifying symbol. The internal numbering code still serves to show the relationship between inputs and outputs. In addition to the coder I/O, there is a general enable which requires a high-level on pin 6 and low-levels on pins 4 and 5 to give a coded output; if EN is not true then all the outputs will be at a high level. If EN is true, the appropriate octal output is active low and other outputs are high. Thus if EN is true and only BIN inputs A and B are high, the internal number is 3, and only OCT output Y3 is low.

Part (c) of the figure depicts an 8-line to 3-line priority encoder; HPRI in the general qualifying symbol signifies 'highest priority', and BIN signifies the binary output. The binary output is obtained from pins 9, 7 and 6 in the indicated weighting (i.e. 1, 2 and 4 respectively). The highest input label number (which is 7) represents the highest priority input, and the lowest input label number (which is 0) represents the lowest priority; the binary output indicates the number (in binary) of the highest priority low-level input. Thus if the input on pin 2 is low, this gives rise to an internal number of 5, providing that the higher priority inputs (on pins 3 and 4) are high; it does not matter what condition the lower priority inputs are in. An internal number of 5 gives rise to low-levels on outputs A2 and A0; thus the highest priority is coded. Notice the interconnection (Z) dependency used to provide inputs to an OR gate from all of the priority inputs to provide a low-level output (GS) providing the E1 input is at a low-level and any one of the priority inputs is at a low level (remember that a low-level at an external input produces a true condition internally for that input when there is an external polarity indicator). The same external input (pin 5) which provides the previously mentioned enable provides a V18 affecting input when at a high-level; this forces a high output at EO even if there is no priority input that is active; any active (i.e. low) priority input also produces a high-level output at EO.

Part (d) of the figure is an example of how the new logic symbols begin to get complicated for some devices: this is a type 7447 BCD to 7-segment decoder/driver; the binary input is applied to the A, B, C and D inputs and the 7-segment display driver lines are those labelled, a,b,c,d,e,f,g (where each lower-case letter represents a particular bar of the display). The T1 note beneath the general qualifying symbol provides a cross-reference to what should be a full font table in order to fully define the output code; this is not represented in the example. The triangular symbol within the general qualifying symbol indicates that the device is a buffer/driver.

Pin 4 of the 7447 device is a wire-AND terminal serving as either a blanking input ($\overline{BI}$) or a ripple-blanking output ($\overline{RBO}$); this pin is both the output of an AND gate or effectively the input to an OR gate; the AND gate output OR the blanking input ($\overline{BI}$ low) gives G21 false; G21 false inhibits the outputs. Each output label has the form 20, 21 and a subsequent passive pull-up output symbol (the display diodes are part of the pull-up network); the 20 indicates an OR relationship with the lamp test input ($\overline{LT}$) such that a low-level on pin 3 OR the appropriate code output enables the segment drives (providing 21 is true due to G21 being true). G21 provides blanking if there is a blanking input (via pin 4), or if there is a zero coded input (CT=0) with lamp test false (pin 3 high) and the ripple-blanking input true (pin 5 low).

Embedded coders

The coder can be useful embedded within a larger symbol in order to create affecting inputs from a coded (or multiplexed) input. *Figure 12.7* illustrates this principle. In part (a) of the figure two inputs become four mode control affecting inputs; compare this

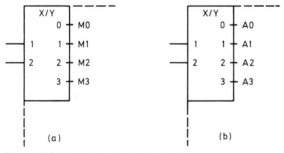

Figure 12.7. Examples of embedded coders

with the bit grouping symbol technique employed in *Figure 7.17* to create the same effect (see page 7–12). Similarly, part (b) of the figure shows how an embedded coder provides an alternative form where two lines give an address dependency; compare this with the method used in *Figure 7.19* (on page 7–15).

13
Signal level converters

This brief chapter may serve as a breather before more complex elements are discussed. It usefully follows the chapter on coders because the general qualifying symbol is the same (*see Figure 12.1*). It deals with signal level converters.

There are many varieties of modern logic devices and inevitably it is necessary to interface between differing signal levels. *Figure 13.1* illustrates how this is achieved. Part (a) shows an ECL to TTL signal level converter (e.g. part of MC 10125). Part (b) shows a dual TTL to MOS signal level converter with common ANDed enables (SN 75365). The levels are represented by the appropriate indication within the general qualifying symbol.

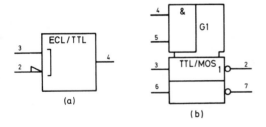

Figure 13.1. Two examples of signal level converters

Although not part of the IEC standard on binary logic elements, it is worthwhile pointing out at this stage that Chapter 20 is a general chapter on *changers*; this chapter has been included to assist the user to minimize symbols needed to show interfacing to logic circuits.

14
Selectors

Selector elements are combinative, and comprise multiplexers and demultiplexers. Multiplexers allow a given number of input lines to be reduced to a smaller number of output lines or to a single line by selection; demultiplexers allow a single line or a given number

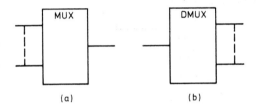

Figure 14.1. General symbols for multiplexer (a) and demultiplexer (b)

of input lines to be routed out to a larger number of output lines by selection. The general symbols for the two forms are shown in *Figure 14.1*, where MUX signifies a multiplexer and DMUX a demultiplexer; the abbreviation DX may be used for a demultiplexer where no confusion can arise.

Multiplexers

Two examples of multiplexers are given in *Figure 14.2*. The device shown in part (a) allows any one of the eight data input lines (D0–D7) to be selected by a binary code applied to the select inputs (A–C), and the selected input is output at Y, or in inverted form at W. The bit grouping symbol indicates the (binary decoder) association between the select lines, the G dependency indicates an AND relationship with the selected data input, and the $\frac{0}{7}$ portion of the label signifies the selection range of 0...7. Thus a high-level on the A and C inputs with a low-level on the B input gives the binary code for 5, which selects the D5 data input by producing an effective G5 affecting input which is true; thus whatever level appears on the D5 input is output at Y, and in inverted form at W. Note that any output is dependent upon a low at the $\overline{G}$ enabling input.

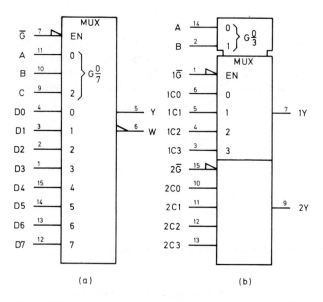

Figure 14.2. Examples of multiplexer (selector) elements: (a) 8-line to 1-line multiplexer (74151) (b) dual 4-line to 1-line multiplexer (74153)

The device shown in part (b) of the figure is a dual 4-line to 1-line multiplexer; it works in exactly the same way as the other example, but because the select lines are in a common control block they select the data inputs in two separate multiplexer elements; each separate element has its own separate enable input. Therefore assuming that both the A and B select inputs are at a high-level, this gives a binary code of 3; this selects the 1C3 and the 2C3 inputs since the effective affecting input is G3. If the enable of the upper element is low then the level at the 1C3 input is identically represented at the 1Y output; if the enable is high, the output at the 1Y element is at a low-level. A similar situation exists for the lower multiplexer element.

Demultiplexers

A simple example of a demultiplexer element is shown in *Figure 14.3*. This shows a 3-line to 8-line demultiplexer. The inputs A, B and C are similar to the multiplexer inputs, and provide an identical G dependency binary-linked affecting input, but instead of selecting an input line, they now produce a low output related to the binary code; this assumes that the AND enable is true, requiring a high-level at the G1 input and low-levels at the two $\overline{G2}$ inputs. Thus if the input condition is A=C=H; B=L (i.e. A and C high and B low), this produces a binary code of 5, giving G5, and causing the Y5 output to go low; all other outputs are high.

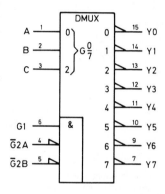

Figure 14.3. A 3-line to 8-line demultiplexer (selector) element (74138)

The element depicted in *Figure 14.3* serves to demonstrate an important principle with the new logic symbols. The symbols are composed to best illustrate a particular logic requirement. The device performing the demultiplexing function in this figure is the 74138; turn back to see that this is precisely the same device which was performing a binary to octal code conversion in *Figure 12.6(b)* (see page 12–5).

If the G1 input is used as a data input, and given that $\overline{G2A}$ and $\overline{G2B}$ are both low, the same device can be used to demultiplex one line to eight lines, controlled by the A, B and C inputs. As an exercise, consider how the symbol might be further modified to better represent this 1-line to 8-line demultiplexer function.

Figure 14.4 shows a dual 1-line to 4-line demultiplexer with an open-circuit L-type output. The A and B lines to the common control block select one of the four possible outputs for each element; the input line in the top element may be taken to be that on pin 1 with pin 2 as an associated active low enable; the lower element is similar except for an inversion on the data input. If either enable input is high the selected output (like the other outputs) goes to a high-level (given an external pull-up resistor), i.e. no outputs are low. Note that the lower element of this two array element repeats the labelling information of the upper element only because the polarity indicators differ on the inputs.

Finally, if you tackled the poser associated with *Figure 14.3* (to convert this to represent a 1-line to 8-line demultiplexer), did you perhaps come up with a solution something like *Figure 14.5*?

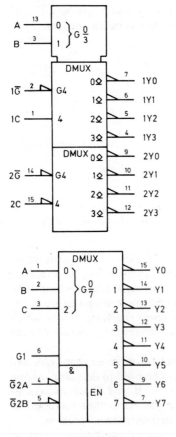

Figure 14.4. Dual 1-line to 4-line demultiplexer (selector) element (74156)

Figure 14.5. A 1-line to 8-line demultiplexer

15
Memory

Memory (or *storage*) elements are ideally suited to the style of representation that the new logic symbology offers, for it enables multi-bit devices to be shown with clarity and ease. The general symbol for a memory element is depicted in *Figure 15.1*. It may be seen that the general qualifying symbol has three distinct sections to specify in order:

(a) the memory type, e.g. RAM, ROM, PROM, CAM (contents addressable memory, associative memory), etc;
(b) the number of addressable locations, i.e. 'words' or 'bytes';
(c) the number of bits to each word.

For anyone not familiar with memory applications, the symbol 'K' is taken to mean 1024 (i.e. the lowest power of 2 above 1000); thus 2K represents $2 \times 1024 = 2048$.

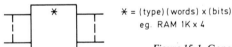

$*$ = (type)(words) x (bits)
eg. RAM 1K x 4

Figure 15.1. General symbol for memory element

Part (a) of *Figure 15.2* depicts a read-only memory of 32-words × 8-bits; it is represented in the general form for a 'register-type' device, utilizing a common control block to represent the five address lines and the common enable. The A against the output of the uppermost element signifies address dependency which is applicable equally to all output lines; thus one of 32 possible address inputs causes a particular word to be selected, and the contents of this address are output via open-circuit L-type outputs when the enable is true (pin 15 low). Part (b) of the figure shows an acceptable version of the symbol which is more compact: it is a recognized form which dispenses with the common control block; it does so at the expense of a rather more tedious task in labelling all the outputs, since there can be no general rule about outputs in

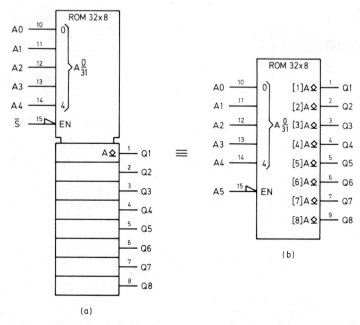

Figure 15.2. Alternative methods of representing a memory element with inputs and outputs (example shows a 32-word × 8-bit ROM, e.g. 7488)

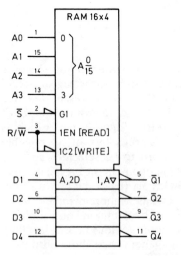

Figure 15.3. A 16-word × 4-bit RAM (e.g. 74189)

a general block. The lowest bit number is nearest the common control block by convention in part (a); the numbers in brackets serve to indicate the order of bits in part (b), where again no such rule can be said to apply.

The choice is purely user preference, although the purist will go for the version with a common control block – this is more in the spirit of the new logic symbology and is a form acceptable for *every* type of memory – which the simple rectangular symbol is not.

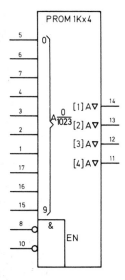

Figure 15.4. A 1K-word × 4-bit PROM (e.g. Intel 3625)

As soon as we get away from the symbol type with a common control block we lose the input/output relationship of separate elements. So any device which has an input (e.g. random access memory) is not so easily represented. In such cases the only clear solution is to use a symbol with a common control block, as shown in *Figure 15.3*. This depicts a 16-word × 4-bit RAM, with address dependency on inputs and outputs. Pin 3 is a READ/WRITE mode selector which operates in conjunction (AND dependency) with an active low chip enable on pin 2; data is entered via pins 4, 6, 10 and 12 in the write mode, or is output via pins 5, 7, 9 and 11 in the read mode. The device has 3-state outputs.

Figure 15.4 shows a 1K-word × 4-bit PROM (programmable read-only memory); only the pins required for normal use are shown (i.e. no programming pins). Because there is no data input to a PROM it can be represented in the simplified form shown.

16

Arithmetic elements

Arithmetic elements form a specialized group of elements which in general do not differ greatly in form from symbols that have been used previously. *Figure 16.1* shows the general symbol for an arithmetic element, where it may be seen that the general qualifying symbol specifies the type of arithmetic element. There is little that can be said about these elements bearing in mind that this book concentrates on symbology rather than function;

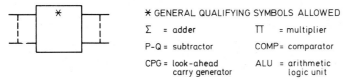

✳ GENERAL QUALIFYING SYMBOLS ALLOWED

Σ   = adder   π   = multiplier

P-Q = subtractor   COMP= comparator

CPG = look-ahead   ALU = arithmetic
      carry generator   logic unit

Figure 16.1. General symbol for an arithmetic element

arithmetic elements tend to be either so simple that an understanding of the function leads to a direct understanding of the element, or so complex that a separate table is required, and such detail would clutter the book: it can be found in the appropriate device data sheets. For this reason the main objective of the chapter is to provide examples of the various types of arithmetic element which are specified in *Figure 16.1*.

Adders and subtractors

Adders come in two basic forms: half-adders and full-adders, where the latter is really no more than two linked half-adders. *Figure 16.2* shows the symbols used for both forms, where the label CO indicates 'carry out' and the label CI indicates 'carry in'.

Figure 16.3 shows a more complex symbol which represents an actual device (the 7480). The upper and lower halves form gated A

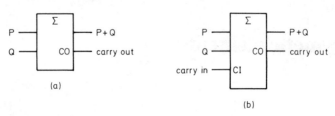

Figure 16.2. Symbols for half-adder (a) and full adder (b)

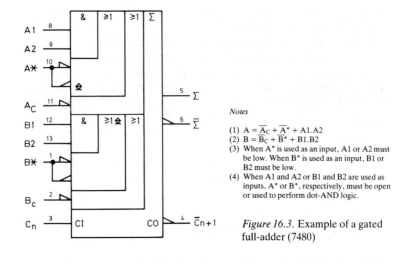

Notes

(1) $A = \overline{A}_C + \overline{A}^* + A1.A2$
(2) $B = \overline{B}_C + \overline{B}^* + B1.B2$
(3) When A* is used as an input, A1 or A2 must be low. When B* is used as an input, B1 or B2 must be low.
(4) When A1 and A2 or B1 and B2 are used as inputs, A* or B*, respectively, must be open or used to perform dot-AND logic.

Figure 16.3. Example of a gated full-adder (7480)

and B inputs to the full-adder, and the normal CI and CO labels are seen at the bottom of the symbol. The only complication worthy of explanation is the A* and B* inputs/outputs: these may be used as an output ($\overline{A1.A2}$ or $\overline{B1.B2}$ respectively), or as an input, providing that the appropriate conditions mentioned in the notes are met. Both true and false (i.e. complementary) outputs are provided, and the variety of input conditions offered makes the device useful in many different configurations (note the OR facility on the actual adder inputs).

Finally two further examples, one of an adder and one of a subtractor. *Figure 16.4* depicts a 4-bit full-adder; note the bit grouping symbols which provide the necessary link between input and output to give bit magnitude. *Figure 16.5* depicts a subtractor, where the inputs are shown to a common control block, and either form of logic levels are allowed for in the two elements below; BI indicates 'borrow in' and BO indicates 'borrow out'.

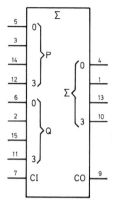

Figure 16.4. A 4-bit full adder (74283)

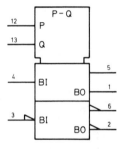

Figure 16.5. A subtractor (MC 1221)

Carry look-ahead generator, multiplier, comparator and ALU

This final section of the chapter provides examples of the more complex functions which are presented in *Figure 16.6* as follows:

(a) carry look-ahead represented by a 4-bit device;
(b) a multiplier represented by a 4-bit parallel multiplier;
(c) a comparator represented by a 4-bit magnitude comparator;
(d) an arithmetic logic unit (ALU) represented by a 4-bit ALU.

It is beyond the scope of this book to enlarge upon the function of such devices, therefore only a few observations are made; clearly anyone intending to use such devices will need a clear understanding of the arithmetic functions before wishing to employ such symbols.

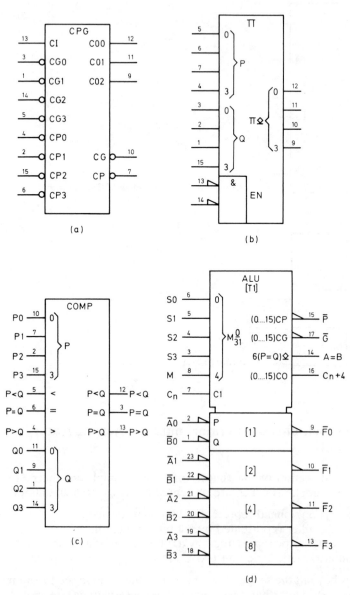

Figure 16.6. Further examples of arithmetic elements: (a) 4-bit look-ahead carry generator (74182) (b) 4-bit parallel multiplier (74285) (c) 4-bit magnitude comparator (7485) (d) arithmetic logic unit (function generator) – a separate table labelled T1 should be appended to detail all functions (74181)

The general qualifying symbol CPG indicates a carry look-ahead generator. Inputs and outputs are as follows:

(a) CG0–CG3 are carry generate inputs;
(b) CP0–CP3 are carry propagate inputs;
(c) CI is a carry in input;
(d) C0–C2 are carry outputs;
(e) CG is a carry generate output;
(f) CP is a carry propagate output.

The general qualifying symbol for a multiplier is the π (pi) symbol. The example shows how bit grouping can relate input and output magnitude.

The general qualifying symbol for a magnitude comparator is COMP; the example shows two 4-bit words being compared (P and Q) with resulting less than (P<Q), equal to (P=Q) and greater than (P>Q) outputs and similar cascading inputs (used when comparing words of greater than 4-bits).

Finally we meet the arithmetic logic unit, or ALU. This performs so many alternative arithmetic functions that it is necessary to append a function table to specify these; this is indicated by the 'T1' in square brackets within the common control block; the table has not been included in this chapter, but may be found in the data sheet for the 74181 device. What function is performed on the associated P and Q inputs of the 4 discrete elements, and hence what the associated outputs will be, is dependent upon the mode set by the four S0–S3 inputs. The M input provides mode control; M is taken high for logic functions or low for arithmetic functions. This device is capable of performing the following functions: addition, subtraction, shift, compare, XOR, AND, NAND, OR, NOR, plus many other special arithmetic functions. Note that the common control block incorporates certain common features relevant to certain functions (i.e. carry in, carry out, carry propagate, carry generate, and equivalence). The bracketed numbers in the elements indicate bit significance.

Part 2
Using the symbols

17

A complex symbol analysed

The point has not been reached in the book where all the binary logic elements described in the IEC standard have been discussed. We can now concern ourselves with how such symbols are used in practice. Chapter 18 considers their use in representing logic circuitry at different levels of treatment, whilst this chapter firstly concentrates on the more complex form of symbol which will be met in manufacturer's data sheets. For numerous examples, the reader is referred to current editions of *The TTL Data Book for Design Engineers* published by Texas Instruments, a company who have helped pioneer the general acceptance of the IEC standard.

It should be borne in' mind that the objective of the semiconductor manufacturer and the objective of the engineer or technical publisher are somewhat different: the semiconductor manufacturer is seeking the most concise symbol which will represent a device; the engineer or technical publisher should be more concerned with the clearest form of representation. Thus the semiconductor manufacturer tends to condense his symbols as much as possible, sometimes at the expense of the user's comprehension. Having said this in a form which might be taken as criticism, I should like to add that there is a genuine case to justify this: they were able to use a simple rectangular symbol to *represent* a complex device before, so why should they not represent *and define* the same complex device by means of a single (but complex) symbol in the new logic symbology? It does mean that it is possible to incorporate that symbol in a circuit diagram intended to show full pin details of all devices whilst at the same time portraying far more information than ever before.

To take this just a little bit further, the use of a complex symbol on a circuit diagram makes matters no worse than using an old style symbol; the user would need to turn to the data sheet for clarification for any devices he was not familiar with in any case. At least he now has more chance of understanding the function of an unknown device than he ever did before, and in the majority of cases, there should be few problems.

What I would like to see is a two-level form of representation by manufacturers for really complex devices. In this chapter we shall consider just one complex device as an example: the 74690 4-bit synchronous counter with output registers and multiplexed 3-state outputs. This title alone tells us that there are three primary separate functions within the device, and it is not surprising, therefore, that the published complex symbol is complicated, as a glance at *Figure 17.1* will show.

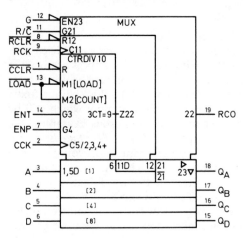

Figure 17.1. Example of a complex symbol for a 4-bit synchronous counter with output registers and multiplexed 3-state outputs (74690)

The IEC guidelines allow the user to embed as much as he likes within a symbol. Here it may be seen that two inner common control blocks are embedded within the outer common control block. This is nothing if not a little intimidating to someone trying to find his feet with the new symbols. The inner common control block represents a decade counter incorporating a common reset (on pin 1); the next larger common control block represents the internal output registers with an independent register clear (on pin 8); the outer common control block represents the final multiplexer which has as inputs the outputs from the output registers and the outputs from the decade counter, and incorporates a 3-state output enable (on pin 12).

All this can be depicted rather more clearly – if taking up rather more space – if it is expanded to show the three primary functions as separate elements in a *detached representation*, as shown in *Figure 17.2.* To aid comparison between the two, the same label numbers have been used. Here the three separate elements of decade counter, register and multiplexer can be clearly seen, and direct comparison shows how the condensed symbol is achieved. The following explanation of the device's function is applicable to either figure since identical labels have been employed.

The counter-divider is a decade counter (signified by the '10' after the 'CTRDIV' portion of the general qualifying symbol). The input on pin 13 selects mode (high for count or low for load). Pin 1 is an active-low counter-clear, and the ripple-carry output on pin 19 goes high at a count of 9 (CT=9), subject to ANDing with input ENT. Parallel loading is performed at the A, B, C, D inputs when pin 13 is low (Mode 1), in response to a clock pulse at pin 2. Pin 2 has an alternative function signified by that portion of the label to the right of the solidi: the counter is clocked by a pulse at pin 2 provided that pin 13 is high (Mode 2) and both enables (G3 and G4) are high. The '+' sign signifies an 'up' counter.

The 4-bit D-type register is used to staticise the output from the counter when clocked by a positive-going pulse applied to pin 9. This register may be independently cleared by a low on pin 8.

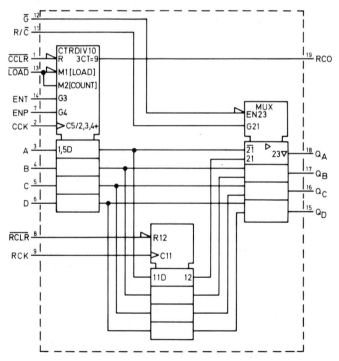

Figure 17.2. A simplified representation of the element shown in *Figure 17.1.*

The output multiplexer selects either the output from the register if the select input on pin 11 is high (dependency label 21 true), or the output from the counter if the same pin is low (dependency label 21 false). Three-state outputs (signified by the inverted triangle on the four outputs) is enabled by a low on pin 12 (EN dependency labelled 23).

The above is readily followed with respect to *Figure 17.2*, and if this is firstly understood, the explanation of the symbol in *Figure 17.1* follows by comparison. Note the *Figure 17.1* divides the three separate functions into what amounts to three-vertical columns, and the inputs to each segment of the common control blocks corresponds to those taken to the different common control blocks in *Figure 17.2*. Because of the need to take an output from the inner (CTRDIV10) common control block (to pin 19) it has been necessary to introduce interconnection dependency in*Figure 17.1* (Z22). Particular note should also be taken of the two horizontal sections of the upper element of the array portion of the symbol; this enables the two alternative inputs to the multiplexer to be shown with their respective dependency labels; as explained previously, it is implied that the same rules apply for the lower elements in the array (ie all bits feed a register and an output multiplexer).

In the early days of the new logic symbology, manufacturers are likely to opt for a lower level of representation for such devices in an older style of representation, but it is to be hoped that once the new logic symbols gain widespread understanding and use, the lower level of representation as exemplified by *Figure 17.2* should be included on the data sheets for such devices to aid understanding.

I should like to take this opportunity of reiterating a point that I made at the beginning of this book. Because the new symbology gives the user the scope to devise symbols of nightmarish complexity, it is to be hoped that no one will actively seek to achieve this objective! We must have all come into contact with certain academic minds which might delight in baffling the common man with such complexities; I would suggest that the production of highly complex logic symbols beyond the ken of the average man should be confined to coffee-time doodling. The purpose of a symbol is to clearly represent a function; if it takes people hours to interpret symbols they will just give up, and if such styles are allowed to develop from the outset, there is the very real danger that the standard will not gain widespread usage. The new logic has a great potential, so those that can appreciate this must keep things down to a sensible level.

Where, then, do we indulge in symbols of the complexity shown in *Figure 17.1*? The obvious answer is where a circuit diagram *must* be contained within a limited space and where such a device would have been previously represented by a simple rectangle. On the other hand, if space permits, there is a lot to be said for avoiding embedded common control blocks, and I would advocate this as an objective if comprehension is to be given a fair chance. It is always possible to put a pecked line around an amplified representation as shown in *Figure 17.2*, putting numbers on the inputs and outputs where they cross this hardware boundary.

If you find the symbol shown in *Figure 17.1* fairly easy to follow, could it be because you have also seen the simplified version? Some examples of other complex devices follow without such simplified representations. See what you make of these before deciding!

Examples of complex devices

I hope that the three examples given in *Figure 17.3* serve to show that the new logic symbology may be versatile, but it cannot tell us everything at a glance! I think that you may agree that a companion simplified form of representation for such devices would not go unappreciated by frustrated engineers working against tight deadlines! (In case I've really got you worried at last, I'm pleased to report that Texas Instruments – at least – generally do just that.) Clearly manufacturers will not sell devices if the user cannot understand what they do.

For the sake of completion, you may be interested to know just a little more about the devices shown in *Figure 17.3*. The 4-bit parallel binary accumulator contains two synchronous registers to contain a Word A (REG4) and a Word B (SRG4), an ALU which operates on the outputs from these two registers, and register control logic and 3-state control logic controlled by the enable decodes from inputs on pins 4, 3 and 2. The dual 16 × 4 bit register file has individual read/write and address controls, 3-state buffer outputs to drive bus lines directly, and is intended for multibus and overlapping file operations. The field-programmable logic, fixed-OR arrays device may be programmed by means of blowing fuse links and comprises an AND OR gate array with 3-state outputs.

One final point worth noting is that a really complex device can be more simply represented by combining parallel lines and using a cross-slash to indicate the number of lines so represented. More will be said of this later.

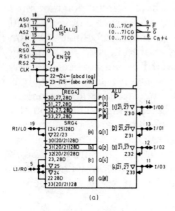

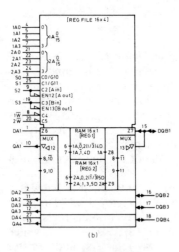

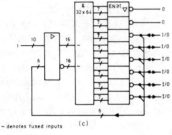

~ denotes fused inputs   (c)

Figure 17.3. Some examples of complex devices: (a) 4-bit parallel binary accumulator (74681) (b) dual 16 × 4 register file (74871) (c) field programmable logic, fixed-OR arrays

Gray boxes

A new feature of the new logic symbols not covered by the first edition of this book is the introduction of the concept of *gray boxes*. This is the method of allowing manufacturers to incorporate a certain amount of information into device symbols which, due to their complex nature, can never completely depict their full internal functions. Obvious cases in the latest generation of devices include field-programmable logic arrays, field-programmable logic sequencers, and, of course, microprocessors! The engineer's beloved *black box* terminology describes the old logic symbols which represented even modestly complex devices by arbitrarily labelled rectangles. The colloquial term *gray box* is now added to describe device representations which are not quite as 'opaque' as the old black boxes; gray boxes do at least provide some useful information.

Examples of gray boxes in use are already to be seen in device manufacturer's data sheets. For example, *Figure 17.4* depicts a 16 × 16-bit multimode multiplier represented as a gray box, and the general qualifying symbol Φ (phi) is used to indicate that it is a gray box.

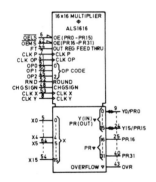

Figure 17.4. An example of a gray box in use (ALS1616 16 × 16-bit multimode multiplier)

In order to counteract the problem of different manufacturers evolving different methods of representing essentially the same device, an international organisation is being set up to standardise device symbols. Clearly it can be expected that it will be a number of years (from the present issue of this book) before this has made a full – international – impact on all manufacturer's data sheets.

18

Different levels of representation

The early chapters of this book considered symbols in general, whilst later chapters tended to concentrate upon specific devices. The only reason to associate the example symbols with specific devices is to aid the reader to interpret the new symbols; a data sheet of a familiar device in a familiar form compared with the same device in its new form can be extremely helpful. This approach also makes the reader familiar with the style of presentation he may expect to find on new data sheets.

A possible disadvantage of consistent reference to actual devices is the expectation that the new symbols can only relate to device level representation. This is not so. The great advantage offered by the new logic symbology is that it can be used at a level of representation to suit the user. This chapter aims to illustrate this principle.

The levels

Chapter 2 defined three terms that it is worth going back to consider:

(a) Pure logic diagram;
(b) Logic diagram;
(c) Circuit diagram.

To summarize these definitions, it may be said that: a *pure logic diagram* depicts logic functions primarily by means of logic symbols in the simplest manner possible: which means that it ignores physical implementation; a *logic diagram* depicts logic functions primarily by means of logic symbols such that the details of all logical relationships are shown, but not necessarily the point-to-point wiring; a *circuit diagram* depicts logic functions and all circuit details in order to show the method of physical implementation.

This chapter considers a hypothetical data processing board and shows how it might be depicted at all three levels of representation.

Three levels – three uses

I believe that the three levels of representation described above may be directly related to three uses, or more pertinently, to three user levels. It leads to an ideal situation in industry, especially for large systems. For example, a large system involving a team of engineers usually breaks down into three levels of engineering: (a) system engineer, (b) design engineer, and (c) electronics engineer. These three levels are concerned with high level system implementation, sub-system implementation and physical methods of implementation respectively. This can be directly related to the three suggested levels of representation as the following table shows.

Engineering level	Design objectives	Level of representation
System engineer	To sub-divide a total system into design modules and to specify signal interfaces	Pure logic diagram
Design engineer	To precisely specify the signal relationships and logic functions without reference to precise method of implementation	Logic diagram
Electronics engineer	To design the circuit at component level	Circuit diagram

It should be made clear that there is no particular aim within the IEC standard to create three specific levels or to relate levels to levels of engineering: this is purely a suggested method of approach that seems to come naturally as a result of the extreme flexibility of the method. Since the system engineer only wishes to consider the system in terms of sub-systems he only wishes to deal in general terms: what would conventionally be block diagram level, but can now become pure logic diagram level. The design engineer is faced with the problem of specifying how the requirements of a given design module should be met, but ideally would like to work at a high level without worrying about what precise components need to be used. The electronics engineer faced with deciding upon the precise method of physical implementation needs an accurate picture of what is required before he can have confidence in meeting the design requirements, and in order that he might specify how the design is to be physically implemented.

An example

We shall now consider an example of the different levels of representation for a hypothetical printed circuit board which is part of a larger system. *Figure 18.1* shows a high level block diagram for a data processing board. Data may be input as 8-bit

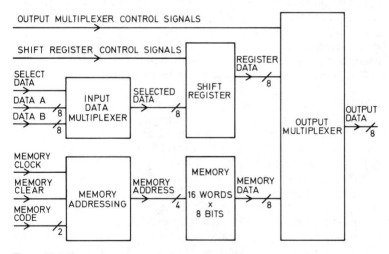

Figure 18.1. Data processing board: block diagram

words from sources A or B; the selected data is then loaded into a shift register for temporary storage, for right-shift, or for right-shift with end-around facility, these facilities controlled by the shift register control signals. An output multiplexer can either select the output from the shift register or the output from a 16-word memory; memory address is decoded from a 2-bit memory code which may be latched into memory addressing logic by a memory clock, or cleared by a memory clear. The block diagram depicts a system engineer's first attempt to define the functions to be performed by this one board within a wider system.

Since the system engineer must rely upon other engineers to design individual boards within his system, he must define all the interface signals to ensure that the system works when the boards are connected together. Thus the block diagram must be refined to include all signal lines, and their precise function must also be defined. *Figure 18.2* shows what might be his next step: the production of a detailed block diagram specifying all signal names. This is how he might have presented it to the design engineer, together with notes specifying the precise requirements of the board. For whilst the detailed block diagram depicts all the signals involved *it does not specify the relationship between signals or the precise logic function of the blocks.*

This is where the new logic symbols come into their own. *This detailed block diagram is now redundant*, and so, to a large degree, are any accompanying notes. For the detailed block diagram may be replaced by a *pure logic diagram* which specifies the complete functioning of the board in terms of the new logic symbology. This is shown in *Figure 18.3*. Note how combined signal lines can be used to great advantage in order to simplify the diagram.

Considering each functional block of the diagram in turn:

(a) The input data multiplexer is simply represented as a multiplexer selecting the 8-bit lines DA0-DA7 or DB0-DB7 according to the input level on ENDA; a high level selects Data A or a low level selects Data B.

(b) The shift register is represented by an 8-bit shift register symbol which specifies the action of all of the shift register control signals. CLREG low gives a reset. SHIFT high specifies a shift mode or LOAD high specifies a parallel load mode from the input multiplexer; CLKREG is the shift register clock. The end-around link is made via the Z4 interconnection dependency and enabled by EAEN.

(c) The memory addressing function is performed by two D-type flip-flops and a code converter; M0 and M1 is a 2-bit memory code which is staticised by clock CKM. This register may be cleared by CLR being taken low. The 2-bit output from the two-element register is converted to a 4-bit memory address by a 2-line to 4-line code converter.

(d) The 4-bit memory address selects one from sixteen addresses within the memory element, thereby producing an 8-bit memory data word.

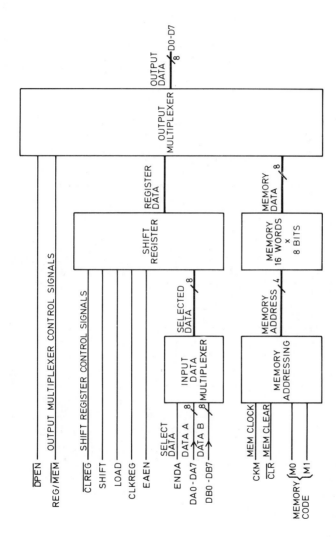

Figure 18.2. Data processing board: detailed block diagram specifying all interface signals

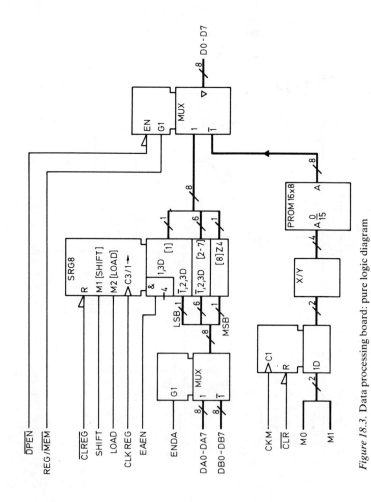

Figure 18.3. Data processing board: pure logic diagram

(e) The output multiplexer selects the shift register output if REG/MEM is high, or the memory output if REG/MEM is low. DPEN is the data processor enable and is used to enable a 3-state output from the output multiplexer (possibly onto an external bus) when low. The 8-bit output is via lines D0–D7.

The following chapter enlarges upon the use of combined lines, but note how similar parallel lines are shown by single lines with a numbered cross-slash indicating the number of lines so represented; note also how branching is easily handled by the numbered slash technique. This particular method of combining lines is not part of the IEC 617–12; it is simply recommended by the author as a standard technique for simplifying pure logic diagrams; it is also a method employed on manufacturer's data sheets.

Close inspection of *Figure 18.3* shows that it fully specifies the function of the data processing board; nothing is left to doubt, and no further explanatory text is required for an engineer who understands the new logic symbols. This is sufficient for him to work with.

The design engineer can now set about drawing up the *logic diagram* which will precisely detail every signal line. The results of his efforts are shown in *Figure 18.4*. Compare this with *Figure 18.3* to see that this is simply a more detailed version of the same design. This is the more conventional form of representation. The foregoing description of operation is naturally equally valid for this figure.

Finally this logic diagram is handed to an electronics engineer responsible for the final implementation of the circuit and he produces a *circuit diagram*; until this stage the design has specified the logic functions to be performed but has left open the way this is achieved. This final stage might be accomplished in an infinite number of ways, some good and some bad; this is where the design skills come in. *Figure 18.5* shows a possible result, although it should be pointed out that this is far from the best design; the particular form shown has been selected because it depicts how different a circuit diagram may be to a logic diagram.

The circuit diagram shows all the devices, complete with pin-out information. Note that AND gates followed by OR gates have been used to implement the input data multiplexer function, whilst a dual 4-bit multiplexer chip is used as the output multiplexer; both perform multiplexer functions and can therefore be represented as multiplexers at the higher level.

Another interesting variation is in the memory. For reasons best known to himself (and his company), the designer has used a

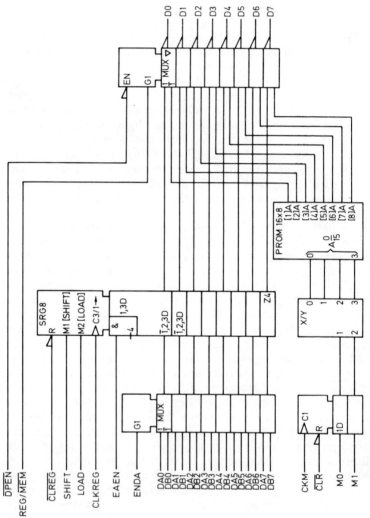

Figure 18.4. Data processing board: logic diagram

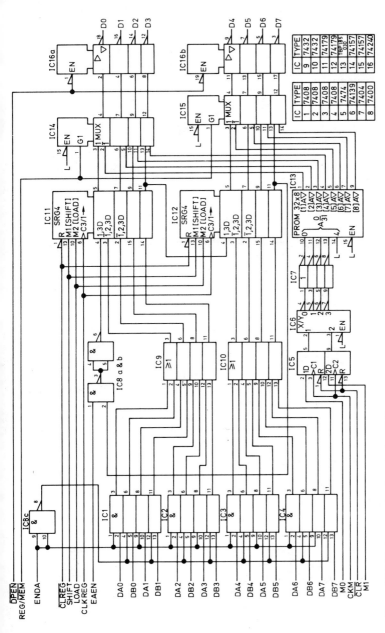

Figure 18.5. Data processing board: circuit diagram

32-word PROM for the memory although a 16-word PROM would suffice; half the PROM is redundant, but because it is used, it must be shown in the circuit diagram. (Such a choice might occur because of better availability, higher reliability, to allow for future expansion, etc.) The device 3-state output is not required.

Instead of an 8-bit shift register, two preferred 4-bit shift registers are employed in the final circuit. Note also that separate gates are required to perform the end-around enable shown so simply at the higher levels by an embedded AND gate and interconnection dependency.

The foregoing example shows just how flexible the new logic symbology is and demonstrates that it introduces a formerly unheard of degree of precision at a high level of representation.

The new logic symbols in documentation

The new logic symbols allow logic functions to be represented at different levels of detail; this is ideal for documentation purposes, since diagrams may be adjusted to suit the needs of the user. The circuit diagram is necessary where the user requires circuit information (e.g. for diagnostic purposes when fault-finding), whilst the logic diagram is adequate when all that is required is to describe the function of a circuit. Similarly, pure logic diagrams may be used at a very high level of representation, when it is only required to give a general idea of the functions performed in the most concise manner possible.

The one major decision to be made before any such diagrams are drawn is whether to use a defined logic convention or to use polarity indicators throughout. Since a positive-logic convention is the most common, many engineers would prefer to see this form of presentation. (The German standard takes this view and excludes polarity indicators.) An isolated diagram which does not employ polarity indicators should always contain a note defining whether a positive or a negative logic convention applies, but in the case of a documentation package, a highlighted statement of which convention is employed should suffice: provided it comes *prior* to any diagrams. The use of polarity indicators, on the other hand, does simplify the problem and no such alternative definitions are required; this is quite unambiguous. If an equipment contains both positive and negative logic conventions then polarity indicators are best used throughout to avoid any confusion. Apart from mixed logic conventions such as this, the choice is up to the originator.

The new logic symbols in commercial organizations

If many different engineers are going to contribute to the production of logic and circuit diagrams, it would appear to be sensible to specify a single method, i.e. positive logic, negative logic, or the polarity indicator convention. The latter is probably the best choice, since then no confusion can arise from the omission of a note about logic convention. It is the most likely choice on data sheets for devices.

Part 3
Closely related symbols

19

Multiple signal paths

This chapter goes beyond the recommendations of the IEC standard, but it does so in order to fill an important gap. It is frequently required to represent a number of signal lines by a single line on a pure logic diagram; it is also sometimes useful to be able to use the technique combined with branching to simplify a complex logic diagram or circuit diagram. In the absence of a definition within the standard on binary logic elements, this chapter aims to provide a few simple ground rules which should enable users to maintain a consistent and unambiguous standard.

The clarity of a diagram can be greatly enhanced by line grouping where there is a definite association between grouped lines, and this is particularly the case for highways or bus lines.

Signal lines

Firstly something should be said about signal lines in general. *Figure 19.1* shows the normal convention for signal lines in part (a) of the diagram. Normal signal flow is from left to right, or from top to bottom, and no arrows *need* be appended for flow in these directions; it is optional to add arrows in such cases. Where the flow is from right to left or from bottom to top then arrows *should* be appended, as shown (unless polarity indicators provide this information). Such lines may be used throughout a diagram.

If it is desired to distinguish data paths or information transfer paths from less important signal lines, such as control signals, then bold lines may be employed. There is no internationally accepted standard agreement on the use of bold lines. The view of the IEC is that bold lines may be used to add emphasis at the discretion of the origination of a diagram.

I believe it is useful to employ bold lines and therefore can do no more than make my own recommendations in this chapter, explaining my reasoning. This technique is sometimes also used to indicate multiple lines; this can lead to confusion and is not recommended. The problem with introducing two types of line is to clearly define what type of signal should be shown in what form; some signals seem to defy definition by falling between control signals and data. For example, the memory code (M0 and M1) and memory address lines in *Figure 18.3* fall between these definitions: they are not pure control signals yet they are not really data.

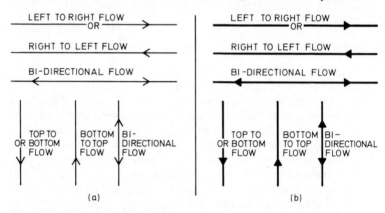

(a)   (b)

Figure 19.1. Varieties of signal flow lines: (a) control signals – may be used for data paths and information transfer if no bold lines used (b) data paths or information transfer. The distinction of bold lines is optional; once introduced into a diagram it must be consistent

In order to get round this problem it is proposed that bold lines are used to represent only:

(a) *data paths*, i.e. signal lines used to transfer 'bits' of data;
(b) *information transfer lines*, i.e. variable control codes that convey *information* but cannot be classified as pure data.

The above definition is likely to lead to less problems than one which attempts to distinguish between major and minor signals; the latter is used quite frequently, but it does present problems since the question of relative importance is a subjective judgement. The recommended definition clearly defines the questioned lines in *Figure 18.3*; M0 and M1, together with the lines into and out of the code converter are information transfer lines, hence they are drawn bold. Such lines are best thought of as transferring variable control codes.

Grouped signals

The recommended symbol for grouped lines is shown in parts (a) and (b) of *Figure 19.2*. It is a familiar method to most engineers already. A cross slash indicates multiple lines, and an adjacent number indicates the number of lines represented by the single line. Part (a) shows the general case, and part (b) shows the equivalent on bold lines. Where it is necessary to depict branching of the multiple lines, an interface line should be used at right-angles to the signal flow, as shown in part (c) of the figure for the general case. This shows a single input line, a group of four input lines, and a group of three input lines, becoming a group of eight lines, which subsequently branch again into a group of two lines, two single lines, and a group of four lines. Clearly the sum of

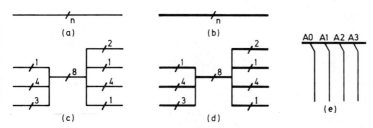

Figure 19.2. Grouped signals, in isolation and branching: (a) 'n' signals grouped (b) 'n' signal lines within a single data path grouped (b) 'n' signal lines within a single data path grouped (c) branching signals (d) branching signal lines within a single data path (note that the interface lines are not bold) (e) individual lines branching from a multiple line

the lines represented must not change anywhere on either side of an interface, i.e. $1 + 4 + 3 = 8$ on the input side, and $2 + 1 + 4 + 1 = 8$ on the output side. Part (d) of the figure depicts the same situation for bold lines, and highlights one important difference: the interface line should not be bold since this is not a signal line in itself.

Bus interconnections

The general symbol for a bus is a broad arrow, as shown in part (a) of *Figure 19.3(b),* where arrows are used at the extremity of the bus; part (b) shows an alternative system employing dot shading within the bus and embedded arrows. The problems only arise at the input or output interfaces with the bus.

Figure 19.3. General symbol for a bus: (a) with end arrows, (b) with embedded arrows and dot shading

Figure 19.4 illustrates the problems which can arise by ambiguous use of the broad arrows. Part (a) of the figure depicts three units interconnected by a bus. It is clear that the bus is an input to Unit A, but what is it to Units B and C? The fact is that documentation frequently uses this form of representation for both outputs and bi-directional connections.

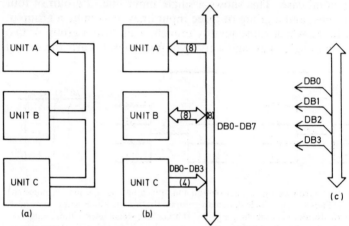

Figure 19.4. Bus interconnections (the bracketed number of lines is optional): (a) ambiguous method (b) unambiguous method (c) representing individual branching lines from a bus

Part (b) of *Figure 19.4* shows the same unit interconnections drawn in the recommended unambiguous manner. Now all becomes clear. Unit A does receive the bus as an input. Unit B is a bi-directional connection to the bus (an input/output), whilst Unit C is an output onto the bus. The rule of thumb is that arrows are used at every bus interconnection except where they can *only* be outgoing.

It is recommended to show unterminated bus lines with arrows, although it is equally acceptable to use blank ends; the advantage of arrows is that the bus format is instantly recognizable.

If it is required to show the number of lines on such a bus, a number is recommended within the bus, as shown bracketed in part (b); the figure shows how this can clarify the situation where fewer than the maximum number of lines are used in a given situation, such as the four least-significant bits only being output from Unit C. This can be further clarified by the addition of signal names, as shown. (The brackets are optional – but consistency should be maintained.) Part (c) of the figure shows how individual lines may be shown branching from a bus.

20
The changer symbol

It is often necessary to shown an interface element, particularly as an input or output to a digital circuit. Although the *changer* symbol is not part of IEC 617-12 on Binary Logic Elements, it is embodied within another IEC publication and features in the national standards of other countries. It is therefore informative to familiarize the reader with its form.

The changer symbol consists of a square sub-divided into two triangles by one diagonal line. On either side of this line is entered a label or symbol to represent the type of signal present on that side of the interface. *Figure 20.1* provides a number of examples.

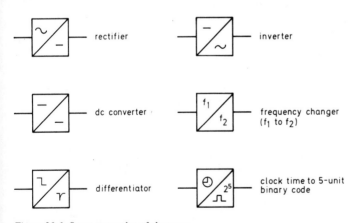

Figure 20.1. Some examples of changers

The rectifier is represented by an a.c. symbol on the input and a d.c. level on the output; the inverter (d.c. to a.c. changer) is the complement of this. A d.c. to d.c. converter therefore comprises a d.c. level in both sections.

A frequency changer is represented by input and output frequency labels stating frequencies. A differentiator may be shown as a square pulse edge transformed into a differentiated pulse. A clock time to binary changer actually depicts a clock face.

It may therefore be seen that the number of variations on this theme is almost limitless. Any sensible symbol or label in the

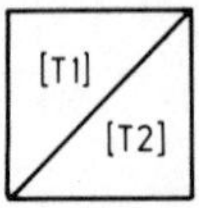

Figure 20.2. A suggested symbol for a changer requiring associated tabular information. T1 and T2 refer to a table detailing the interface

changer should prove acceptable where no other symbol fits an interface situation. If the interface is really complex, I can only suggest the application of a technique met before within the standard for logic symbols. *Figure 20.2* replaces the normal labels in the changer symbol by references to an associated table or tables; ideally a single table with columns T1 and T2 should show the relationship.

Appendix 1

Glossary of symbols and notations

The purpose of this appendix is to provide a quick reference to important topics covered in depth within this book. If insufficient detail is found within the appendix then the main body of the book should be consulted.

Information contained within this appendix is embodied in four tables as follows:

Table 1 – Qualifying symbols for inputs and outputs
Table 2 – Symbols used inside the outline
Table 3 – General qualifying symbols
Table 4 – Dependency notation

TABLE 1.   **Qualifying symbols for inputs and outputs**

Signal flow

Signal flow from right to left (otherwise left to right assumed).

Bidirectional signal flow.

Active low

Active low input. Equivalent to ──◁ in positive logic.

Active low input with right-to-left signal flow.

Active low output. Equivalent to ▷── in positive logic.

Active low output with right-to-left signal flow.

Logic negation

Logic negation at input.

Logic negation at output.

Dynamic inputs – active on transition indicated in table

Positive logic	Negative logic	Polarity indication
$0 \rfloor^{1}$	$0 \rceil_{1}$	$L \rfloor^{H}$
$1 \rceil_{0}$	$1 \rfloor^{0}$	*N/A*
N/A	*N/A*	$H \rceil_{L}$

Non-logic connections

Defined by label inside symbol (e.g. power supply connection).

Analogue signal input.

Virtual connections

Virtual (internal) input stands at internal 1-state unless affected by an overriding dependency relationship.

Virtual (internal) output. Its effects on an internal input to which it is connected must be indicated by dependency notation.

Internal connections

Non-inverting internal connection.

Inverting internal connection.

Dynamic internal connection.

TABLE 2.   **Symbols used inside the outline**

Miscellaneous inputs

Bi-threshold input (input with hysterisis).

Enable input: 1-state enables all outputs.

Query (interrogate) input: 1-state causes interrogation of the contents of associated element.

Fixed-state input: always stands at its internal 1-state.

Miscellaneous outputs

Postponed output. The output changes when the initiating input returns to its initial external state or level.

Open-circuit output. One of the two possible internal logic states corresponds to a high impedance condition; in order to establish a valid logic level in this condition an external pull-up or pull-down is required. Where it is required to indicate which state is high impedance, one of the following two symbols should be used. Distributed connection possible.

Open-circuit L-type output. When not in its high impedance state the output produces a low impedance low level. Requires external pull-up circuit. Distributed connection possible.

Open-circuit H-type output. When not in its high impedance state the output produces a low impedance high level. Requires external pull-down circuit. Distributed connection possible.

Passive pull-up output. Similar to open-circuit L-type output but containing internal pull-up resistor. Distributed connection possible.

Passive pull-down output. Similar to open-circuit H-type output but containing internal pull-down resistor. Distributed connection possible.

3-state output. This output has a third external high impedance state which has no logic significance.

Fixed-state output: always stands at its internal 1-state.

Extension

Extension input. An input to which the output of an extender element may be connected.

Extender output. An output which may be connected to the extension input of another binary element in order to extend the number of inputs of that element.

TABLE 2.   **Symbols used inside the outline (continued)**

Bistables

—[S]   S input: set. An internal 1-state sets the bistable.

—[R]   R input: reset. An internal 1-state resets the bistable.

—[T]   T input: toggle. An internal 1-state complements the bistable.

—[J]   J input: normal significance with respect to JK bistable.

—[K]   K input: normal significance with respect to JK bistable.

—[D]   Data input: normal significance with respect to D-type bistable.

Shift registers

—[→]   Shift right (down). An internal 1-state causes data to be shifted one position to the right on horizontal symbol or down on a vertical symbol. Vertical symbol preferred.

—[←]   Shift left (up). An internal 1-state causes data to be shifted one position to the left on horizontal symbol or up on a vertical symbol. Vertical symbol preferred.

Counters

—[+]   Count up. An internal 1-state causes the count of the element to be increased by one.

—[−]   Count down. An internal 1-state causes the count of the element to be decreased by one.

Content

—[CT=m]   Content input. Replace m by an appropriate indication of the content of the element resulting from taking this input to an internal 1-state. Used for loading a register.

[CT=m]—   Content output. Replace m by an appropriate indication of the content of the element when the output is at an internal 1-state. Used for detecting a count.

Comparators

—[>]   Greater-than input, for use when representing cascaded comparators.

—[<]   Less-than input, for use when representing cascaded comparators.

—[=]   Equal input, for use when representing cascaded comparators.

[*>*]—   Greater-than output. Each * must be replaced by the designation of an operand.

TABLE 2.   **Symbols used inside the outline (continued)**

$\boxed{*<*}-$	Less-than output. Each * must be replaced by the designation of an operand.
$\boxed{*=*}-$	Equal output. Each * must be replaced by the designation of an operand.
	Note: The symbols $< = >$ may be combined, as in the following example: $\geq$ *meaning 'equal to or greater than'*
$\boxed{!}-$	Compare (match) output of an associative memory.

Arithmetic devices

$-\boxed{Pm}$	Operand input (P input shown). Preferred letters for operands are P and Q; where these letters are not suitable or more than two operands are involved, other characters may be employed providing no confusion may arise. The m is replaced by a numeral representing the weight of the bit expressed as either the exponent of the power of 2 or as its equivalent decimal value. If an operand comprises 2 or more bits, the binary grouping symbol may be employed (*see below*).

$-\boxed{BI}$ Borrow-in input.

$-\boxed{BG}$ Borrow-generate input.

$\boxed{BG}-$ Borrow-generate output.

$\boxed{BO}-$ Borrow-out output (i.e. ripple)

$\boxed{BP}-$ Borrow-propagate output.

$-\boxed{BP}$ Borrow-propagate input.

$-\boxed{CI}$ Carry-in input.

$-\boxed{CG}$ Carry-generate input.

$\boxed{CO}-$ Carry-out output.

$\boxed{CG}-$ Carry-generate output.

$-\boxed{CP}$ Carry-propagate input.

$\boxed{CP}-$ Carry-propagate output.

A number may be added as a suffix to express the exponent of the power of 2, being the weight of individual bits.

TABLE 2.   **Symbols used inside the outline (continued)**

Binary grouping

Bit grouping symbol for inputs (qualifying symbol for multi-bit input), general symbol. Inputs grouped by this symbol produce a number that is the sum of the individual weights of the inputs standing at their internal 1-states. M_1 to m_k are replaced by the decimal equivalents of the actual weights; if all weights are powers of 2, they may be replaced by the exponents of the powers of 2. Labels between m_1 and m_k may be omitted providing that no confusion can arise. The number may be: a number upon which a mathematical function is to be performed; an identifying number in the sense of dependency notation; a value to become the content of the element. The * is replaced by an indication of the operand on which the mathematical function is to be performed (e.g. P or Q), by an indication in the sense of dependency notation, or by CT; in the latter case the number produced by the inputs is the value of the content that is loaded into the element.

Bit grouping symbol for outputs (qualifying symbol for multi-bit output), general symbol. The general notes above apply with respect to an output, and where the number may be the result of a mathematical function or the value of the content of the element. Where * is replaced by CT, the number represented is the actual value of the content of the element.

Line grouping

Line grouping symbol for inputs. This symbol indicates that two or more terminals are needed to implement a single logic input. See example to right.

Line grouping symbol for outputs. This symbol indicates that two or more terminals are needed to implement a single logic output. See example to right.

TABLE 3.   **General qualifying symbols**

&	AND gate or function.
≥1	OR gate or function.
=1	EXCLUSIVE-OR gate or function. One and only one input must be active to activate the output.
=	Logic identity. All inputs must stand at the same state.
2k	An even number of inputs must be active.
2k+1	An odd number of inputs must be active.
1	The single input must be active (i.e. non-inverting buffer).
▷ or ◁	A buffer or element with greater than usual output driving capability. Symbol points in direction of signal flow.
⎍	Schmitt trigger; element with hysterisis; bi-threshold detector.
X/Y	Coder. A code converter from 'X' to 'Y', where substitutions for X and Y may indicate types of code, e.g. X/DEC, BCD/Y, DEC/BCD, BIN/7-SEG, etc.
MUX	Selector: multiplexer.
DMUX	Selector: demultiplexer. DX may also be used.
Σ	Adder.
P-Q	Subtractor.
CPG	Look-ahead carry generator (carry propagate and generate).
π	Multiplier.
COMP	Comparator.
ALU	Arithmetic Logic Unit.
⎍	Retriggerable monostable.
1⎍	Non-retriggerable monostable.
G ⎍	Astable element; general symbol. A signal generator producing an alternating sequence of 1 and 0. The 'G' is the qualifying symbol in this and astable elements given below; showing the waveform is optional.
!G ⎍	Astable element, synchronously starting.
G! ⎍	Astable element, stopping after completion of last pulse.
!G! ⎍	Astable element, synchronously starting, stopping after completion of last pulse.
SRGm	Shift register, where substitution for m specifies number of bits.
CTRm	Counter, where substitution for m specifies number of bits.

TABLE 3.   **General qualifying symbols (continued)**

CTRDIVm	Counter/divider, where substitution for m specifies cycle length (e.g. CTRDIV10 = decade counter/divider).
ROM	Read only memory.
PROM	Programmable read only memory.
EPROM	Erasable programmable read only memory.
RAM	Random access memory.
DRAM	Dynamic random access memory.
FIFO	First-in first-out memory.
CAM	Content addressable memory.

TABLE 4.   **Dependency notation**

The letter 'm' in this table is used to denote an identifying number; substitution of an appropriate number is required in normal usage. Within a given element, the preferred method is to utilize consecutive numbers starting from '1'.

Gm   AND dependency is denoted by the letter 'G'. Each input or output affected by a Gm input or a Gm output stands in an AND relationship with this Gm input or Gm output.

Vm   OR dependency is denoted by the letter 'V'. Each input or output affected by a Vm input or a Vm output stands in an OR relationship with this Vm input or Vm output.

Nm   NEGATE dependency is denoted by the letter 'N'. Each input or output affected by an Nm input or Nm output stands in an EXCLUSIVE-OR relationship with this Nm input or Nm output.

Zm   INTERCONNECTION dependency is denoted by the letter 'Z'; it is used to indicate the existence of *internal* logic connections between inputs, outputs, internal inputs and/or internal outputs.

Cm   CONTROL dependency is denoted by the letter 'C'; it is used for sequential elements where more than a simple AND relationship is implied. An internal 1-state allows affected inputs to have their normally defined effect on the function of the element; an internal 0-state prevents affected inputs from having any effect on the function of the element.

ENm   ENABLE dependency is denoted by the combination of letters 'EN'; it is used to indicate an ENABLE input which does not necessarily affect all outputs of an element. It can also be used when one or more inputs of an element are affected.

Mm   MODE dependency is denoted by the letter 'M'; it is used to indicate that the effects of particular inputs and outputs of an element depend upon the mode in which the element is operating. When an Mm input or Mm output is at its internal 1-state, affected inputs have their normally defined effect on the function of the element and affected outputs stand at their normally defined internal logic states; when an Mm input or Mm output is at its internal 0-state, the affected inputs have no effect on the function of the element and any sets of labels at any output containing that identifying number shall be ignored. (Note that M-dependency affects inputs in the same way as C-dependency; the distinction is that the letter C should be used to indicate an input which causes action, whereas the letter M is used to indicate an input that has alternative preparatory effects.)

Am   ADDRESS dependency is denoted by the letter 'A'. This dependency allows specific words in an array to be identified by means of the identifying number within an Am input or an Am output. The form 'Am' at an input or output is associated with the form 'A' at an affected input or output. (Thus A1, A2 or A3 are associated with an input or output containing A in the label, although three separate words are represented by that portion of the element containing the A form, assuming that there are only three different Am form inputs or outputs.)

TABLE 4.   **Dependency notation (continued)**

Sm	SET dependency is denoted by the letter 'S'. When an Sm input is at its internal 1-state, affected outputs take on the internal logic state normally associated with the input combination $S=1$, $R=0$, regardless of the state of any R input; when an Sm input is at its internal 0-state it has no effect on the element.
Rm	RESET dependency is denoted by the letter 'R'. When an Rm input is at its internal 1-state, affected outputs take on the internal logic state normally associated with the input combination $S=0$, $R=1$, regardless of the state of any S input; when an Rm input is at its internal 0-state it has no effect on the element.

Appendix 2
Participating countries

Before considering the countries participating in this international standard it is worthwhile giving a brief history of the work which has paved the way: this is no overnight development!

Derivation of the IEC standard

The standard for logic symbols upon which this book is based is the International Electrotechnical Commission (IEC) Publication 617: Graphical Symbols for Diagrams, Part 12: Binary Logic Elements. (IEC Publication 617 replaces IEC Publication 117.) This standard now takes into account the use of computer-aided drafting equipment, and all symbols are designed on a grid; a transparent overlay with this grid is included in IEC Publication 617-1.

The new standard (i.e. the 'new logic symbols' described in this book), was prepared by Sub-Committee 3A: Graphical Symbols for Diagrams, of IEC Technical Committee No. 3: Graphical Symbols.

Draft documents produced by Working Group No. 2 during ten meetings between 1973 and 1979 were discussed at a meeting of Sub-Committee 3A held in The Hague in 1979, and as a result of this meeting, IEC Document 3A (Central Office) 115, was submitted to the National Committees for approval under the Six Month's Rule in April 1980. Although many countries voted in favour of publication, four countries voted against unless certain amendments were made (these countries were Finland, France, the United Kingdom and the United States of America). In accordance with a decision taken at a meeting held in Baden-Baden in September 1980 the document was not published. The working group was asked to consider all comments and to assist the Chairman of Sub-Committee 3A in the preparation of an amended draft; as a result, IEC Document 3A (Central Office) 131 was submitted to the National Committees for approval under the Two Month's Procedure in May 1981. This resulted in the following countries voting explicitly in favour of publication: Australia, Austria, Belgium, Denmark, Egypt, Finland, Germany, Japan, Netherlands, Spain, Sweden, Switzerland, United Kingdom, and the United States of America. (It does not follow that countries not voting were not in favour.)

The first edition of the final IEC Document 617-12 was published late in 1983 and went into general circulation in 1984; this is the document upon which participating countries intend to base their own national standards. In order to promote international unification, the IEC expresses the wish in this document that all National Committees should adopt the text of the IEC recommendation for their national standards in so far as national conditions will permit; any divergence between the IEC recommendation and the corresponding national standards should ideally be clearly identified in the latter.

Degree of conformity

So it came to pass that the deliberations of many countries throughout the world (including interested parties from Soviet countries) over a ten-year period culminated in an international standard; the general conclusion to be drawn is that the majority of the participating countries intend to embody this standard within their own national standards as soon as possible. (In fact certain countries had already done so before publication of IEC 617-12, and others had revised their standards in anticipation of the new style of symbol.)

This book is intended for worldwide circulation and is therefore based upon the IEC recommendations rather than any single national standard. Because of this, the following table is intended to show the degree of national conformity with the IEC standard. To this end the various national bodies were individually approached for the purposes of compiling the table; not surprisingly, not all responded (not all responded to the IEC). Where no definite response was received the *Conformity* column contains an opinion based upon the country's voting responses to the IEC; an asterisk denotes such entries.

The *Notes* column refers to the numbered notes immediately following the table; these are used to provide further details where necessary.

Country	Conformity	Document ref.	Notes (see overleaf)
Australia	general*		4
Austria	general*		4
Belgium	total	NBN C 03-115 (6th Edition) Operateurs logiques binaires	
Canada	general*		5
Denmark	general	DS 5009.12	1
Egypt	general*		4
Finland	total	SFS 4612 Graphical symbols for diagrams: Binary logic elements	
France	total	NFC 03–212 Operateurs logiques binaires	
Germany	general	DIN 40 900, Teil 12 Schaltzeichen; Binare Elemente; 617-12 modifiziert	2
Greece			6
Ireland	none	no standard	
Israel	general*		5
Italy	total	Italian CEI standard	
Japan	general*		4
Luxembourg			6
Netherlands	total	NEN 10 617-12 Logicasymbolen	
Norway	total	NEN 221.84	
Portugal			6
Spain	general*		4
Sweden	total	SS 421 25 08	3
Switzerland	total	SEV/ASE 9617-12	
Turkey	general*		5
United Kingdom	total	BS3939: Part 12 Graphical symbols for electrical power, telecommunications and electronic diagrams: Binary logic elements	
USA	total	ANSI/IEEE Std 91-1984 Graphic Symbols for Logic Functions	4

Notes to the preceding table

(1) Danish standard extended to include three annexes.
(2) German standard excludes polarity indicators.
(3) Swedish standard includes amendment containing earlier standardized symbols.
(4) No specific confirmation on the degree of conformity received, but since the country voted in favour of the publication of the final IEC standard, general conformity is assumed.
(5) No specific confirmation on the degree of conformity received, but since the country voted in favour of the first draft of the IEC standard, and the amendments made were few, it is considered reasonable to assume general conformity.
(6) Intentions of country unknown.

Note concerning conformity

Should the National Committee of any country wish to amend any information relating to their country contained within this appendix they are requested to contact the publishers without delay if they wish such amendments to be incorporated in the next edition.

Appendix 3
Changes in this edition

Introduction

This appendix is provided for the benefit of teachers and lecturers who may also have used the First Edition of this book, or who may need to mix editions. Its purpose is to briefly indicate the principal differences.

Polarity indicator convention

The term *polarity indicator convention* is now used in place of *mixed logic convention*. There is no approved title for this convention in the IEC standard and the term used in the First Edition was one in general usage at that time. Since this term is now deprecated and misleading – polarity indicators imply no logic convention rather than a 'mixed' logic convention – the European-favoured (and highly suitable) term *polarity indicator convention* is now employed.

Note that different countries are liable to use different terms; the USA prefers the term *direct polarity indication,* and the term *mixed logic convention* may still be in general use for a time.

Horizontal shift registers and counters

Chapter 11 originally provided examples of horizontally positioned shift registers (*Figures 11.2 and 11.3*). As explained in this edition on page 11-2, there is far from a clear convention. IEC 617-12 provides no examples and makes no recommendations about this, but other national standards which do make such recommendations do not agree and only give rise to confusion. It is recommended to avoid horizontal placement altogether unless working to a specific national standard.

For information, it may be noted that the simplest rule of thumb for the acceptance of horizontal symbols is to only allow rotation of the otherwise unchanged symbol to give the required vertical signal flow; this implies that labels are also rotated. Different national standards have different ideas on how to go about avoiding sideways labels, and this is where the problems begin!

Other changes

Other changes are minor. Attention is drawn to the following.

1. *Figure 4.1*. Reference to external logic convention has been removed to avoid confusion.

2. *Page 5-6*. It is made clear that the open-wired forms of distributed connection are not part of the IEC standard.

3. *Page 7-12*. Because the original examples of mode dependency affected inputs were not supported by direct IEC examples and the effects may be modified by current discussions such that the original description would be misleading, a more straightforward example has been used as *Figure 7.17(b)*, based upon an IEC example.

4. *Page 7-14*. Because the general term *byte* is now usually taken to imply exactly eight bits, this term has been avoided.

5. *Figure 12.3*. The figure represents a coder in which one output (Y) is active for a given input condition (X). In order to emphasise the fact that only one output was ever active, this was treated as a (partial) decimal decode in the First Edition. Unfortunately, this is a rather misleading area altogether, for this particular symbol is *traditionally represented* as a BIN/OCT decoder. This stems from the fact that it is common to associate the term *octal* with eight (or seven) lines and the term *decimal* with ten (or nine) lines. Whilst this leads to little confusion with most devices (e.g. octal bus buffers), I feel that it is misleading when applied to *coders*, which purport to convert from code X to code Y. Thus to term the device in *Figure 12.3* a BIN/OCT coder would seem to imply conversion from a binary code to an octal code – which it does not – a binary code is simply converted to one of seven lines. Compare this with the 74185 BIN/BCD coder, for example, where this really does convert between binary and binary-coded-decimal codes.

Since this book is to be used for teaching and the definitions for such terminology have not yet been agreed or published, I have reverted to the more basic X/Y general qualifying symbol, since this is quite unambiguous. I feel that coders should only really use the formal meaning of these words which are inherently associated with codes; thus the device in question is more truly an OCT/DEC coder, since an octal code at X produces the decimal equivalent at Y; this does overlook the missing 'zero' condition, however.

This subject is under discussion at the time of writing. I believe that the likely outcome for the terms DEC and OCT will be that these definitions may be associated with either n or n-1 lines (where n is 10 or 8 respectively); in the case of n-1 lines, the nth state is that at which all lines are at a logic 0-state. If this comes to pass, such designations will continue to apply to coders, and will, no doubt, continue to confuse students!

Index

Adder, 16–1
ALU, 16–3
Analogue connections, 4–8
Arithmetic elements, 16–1
Arithmetic logic unit (ALU), 16–3
Arrays, 3–4
Astable elements, 9–6

Binary grouping, 7–15, A1–6
Binary logic element, 2–2
Binary-octal code converter, 12–4
Binary variable, 2–2
Bistable, 2–4, 9–1
  non-volatile, 9–8
  toggle, 9–8
  with initial power-up state, 9–7
Bi-threshold input, 4–7
Buffers, 5–6
Bus interconnections, 19–3

Carry look-ahead generator, 16–5
Changer symbol, 20–1
Circuit diagram, 2–2, 18–7
Closely related symbols, Part 3
Code converter, 12–4
Coder, 12–1
  embedded, 12–7
Codes, input and output, 12–1
Combinative element, 2–4, 5–1
Commercial use of new symbols, 18–11
Common control block, 10–1
Common output element, 3–3
Comparator, 16–5
Complex devices, examples of, 17–5
Complex symbol analysed, 17–1
Content output, 8–3
Control signals, 19–2
Counter, 11–5
Counter/divider, 11–5

Data lockout bistable, 9–3
Data paths, 19–2
Decimal coder, 12–3
Definitions, 2–1
Delay element, 2–4, 6–1
  specified delay, 6–1
  tapped delay elements, 6–2
  unspecified delays, 6–3
  variable delays, 6–3
Demultiplexer, 14–3
Dependency notation, 7–1, A1–9
  Address (A), 7–14
  AND (G), 7–3
  Control (C), 7–10
  Conventions, 7–2
  Enable (EN), 7–11
  Interconnection (Z), 7–7
  Mode (M), 7–12
  Negate (XOR), 7–6
  OR between affecting I/O with
    identical labels, 7–5
  OR (V), 7–6
  Set/Reset (S/R), 7–8
  Summary, 7–17
Direction of information flow, 3–4
Distributed connection, 5–6
Documentation with new symbols,
  18–10
Dot function, 5–6
Dynamic input, 4–5

Enable input, 7–11
Engineering levels, 18–2
Extender output, 4–7
Extension input, 4–7
External logic state, 2–5

Fixed-mode input, 4–9
Fixed-state output, 4–9

Index

Gates, simple types, 5–1
General qualifying symbol, 2–5, 3–1,
    A1–7
Gray boxes, 17–7
Grouped signals, 19–2

Hysteresis:
  gates, 5–6
  input, 4–7

Information transfer, 19–2
Inputs and outputs, 4–1
  (see also under particular types)
Internal input/output, 4–9
Internal logic state, 2–4
I/O abbreviation, 7–2

Label:
  input, 8–1
  output, 8–3
  sequences, 8–1
Latch, 7–9, 9–1
Levels of representation, 18–1
Logic:
  convention, 2–3, 4–1
  diagram, 2–1, 18–7
  function, 2–1
  levels, 2–2
  states, 2–2
  symbol, 2–1
  threshold gates, 5–3

Majority element, 5–4
Memory, 15–1
Mixed logic convention, 2–3
Monostable, 9–4
Multiple signal paths, 19–1
Multiplexer, 14–1
Multiplier, 16–5

Negative logic convention, 2–3, 4–2
Non-logic connections, 4–7
Notations, glossary of, A1–1

Open-circuit outputs, 4–9

Parity elements, 5–4
Participating countries, A2–1
Polarity, 4–1
  indicator convention, 2–3
  indicators, 4–3
Positive logic convention, 2–3, 4–1
Postponed output, 4–6
Priority encoder, 12–6
Pure logic diagram, 2–1, 18–4

Qualifying symbol, 2–5, 3–1

Schmitt trigger inputs/gates – see
    'hysteresis'
Selectors, 14–1
Sequential element, 2–4, 11–1
Shift register, 11–2
Signal level converter, 13–1
Signal lines, 19–1
Special outputs, 4–9
Storage element, 2–4, 15–1
Subtractor, 16–1
Symbols:
  combination of, 3–2
  composition, 3–1
  embedded, 3–3
  glossary of, A1–1
  qualifying, A1–2
  use of, 17–1

Three-state outputs, 4–9
Two-state elements, 9–1

Virtual input/output, 4–9

Wired function, 5–6

Supporting slide package

With the introduction and growing acceptance
throughout the world of the new logic symbols,
industry now needs engineers who are fully familiar
with them. For many companies this will mean the
retraining of existing digital and microprocessor
engineers and specific training for those joining.

To meet this need a slide package is being made
available for training in industry to support this
book, comprising

* A lecturer's copy of *A Practical Introduction to the
  New Logic Symbols 2ed*
* Explanatory material
* Slide summary
* Approximately 70 slides for use on overhead
  projectors
* Discount on additional copies of the book for
  students use

when purchased with the slide package

For further details and information on the price,
please contact:

**The Electronics Engineering Editor, Butterworth
Scientific Ltd, PO Box 63, Westbury House, Bury
Street, Guildford, Surrey GU2 5BH, England**

microprocessors
and microsystems

**the authoritative international journal
on microcomputer technology and
applications for designers**

Special individual subscription rate

Further details and subscription rates from:
**Sheila King Butterworth Scientific Limited
PO Box 63 Westbury House Bury Street Guildford
Surrey GU2 5BH England**
Telephone: 0483 31261   Telex: 859556 SCITEC G

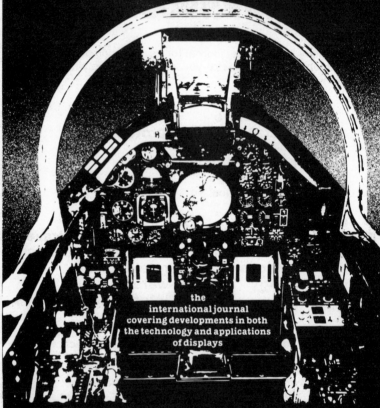

DISPLAYS
Technology and Applications

the
international journal
covering developments in both
the technology and applications
of displays

Published four times a year by Butterworth Scientific Limited
P.O. Box 63 Westbury House Bury Street Guildford Surrey GU2 5BH England
Telephone: 0483 31261 Telex 859556 Scitec G